Physics and the Sound of Music

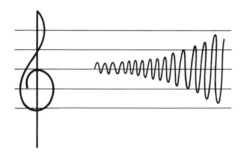

John S. Rigden
University of Missouri—St. Louis

John Wiley & Sons New York Santa Barbara London Sydney Toronto

Library of Congress Cataloging in Publication Data

Rigden, John S.
 Physics and the sound of music.

 Includes index.
 1. Music—Acoustics and physics. I. Title.
ML3805.R56 781'.1 77-5638
ISBN 0-471-02433-3

Printed in the United States of America

10 9 8 7 6 5 4 3 2 1

Physics and the Sound of Music

Preface

In the last few years, the subject of musical acoustics has grown in popularity. Many universities and colleges now offer courses in this subject area. These courses serve not only music students but they appeal to a broad cross section of student interests. The subject is interdisciplinary, as it embraces physics, music, psychology, physiology, neurology, and architecture. This book is written for students who want to learn about musical sound.

About 2500 years ago Pythagoras discovered that the pitch of a sound depends on the length of the string that produces it. Moreover, he found that two similar strings stretched with equal tautness produce pitches corresponding to an octave when the second string is *exactly* twice the length of the first. Other harmonious musical intervals result when the length ratio of the two strings equals small whole numbers.

The simple whole-number exactitude attending the harmonious quality of the musical intervals so inspired Pythagoras that he dreamed of finding in the numbers of musical harmony the laws governing the motions of the stars. Twenty-five centuries have elapsed since his time, but the same dream, expressed in different idiomatic styles, has excited the thoughts of men and women in every era.

The subject of this book is a direct descendent of that epochal discovery made by the man from Samos, Pythagoras.

Physics and the Sound of Music covers a range of topics: from the vibrating string to the factors that contribute to the expressive eloquence of a Stradivarius; from a vibrating air column to the problems inherent in the valve systems of brass instruments; from the reflection of sound waves to the characteristics of an environment praised for its "good acoustics"; from the human auditory system to the subjective and objective dimensions of consonance and dissonance; from transducers to high-fidelity sound systems; from pure tones to electronic synthesizers. In the final analysis, however, this book is about the production, the propagation, and the perception of that sound we call music.

Music is produced by a variety of instruments. The stringed instruments, the woodwinds, and the brass instruments are all quite different; yet, the same few acoustical principles enable us to understand the means by which these instruments produce their characteristic sounds. While musical instruments are the sources of musical sound, these instruments are played in an acoustical environment. As the sound propagates away from the source, the surrounding environment actively participates in shaping both the intensity and the quality of the perceived sound. If the shaping process enhances the perceived sound, the environment is acclaimed for its acoustical qualities. When a sound wave enters the auditory canal and starts the eardrum vibrating, a chain of events is initiated that leads to perception. The human auditory system further shapes the sound.

But we need not go a concert hall to hear fine music. Today, more than ever before, there is a pulse to our living spaces that is created by the beat and rhythm of music. During the last decade quality high-fidelity sound equipment has come within reach of most budgets with the result that stereo sound systems occupy prominent positions in our home, family rooms, and in our vehicular dashboards. Furthermore, FM stations, record albums, and tape-recordings provide such a variety of program material that any musical taste can feel satisfied. The steps leading to the purchase of a high-fidelity system can be frustrating and confusing as the options confronting the buyer increase. The material in this book should not only enable the listener to understand something about the system that provides such enjoyable experiences, it should also help the consumer as the purchase decision is consummated.

This book is written for the individual who enjoys music and whose interest and curiosity extends to the means by which musical sound is produced and reproduced. Although the subject is technical, no specific technical background is assumed. A reader who is familiar with the science and mathematics typically studied in high school should have no difficulty understanding the material contained in the following pages. Some musical background such as piano lessons would be useful; however, it is not essential.

I enjoy the subject presented in this book. I hope the reader enjoys it too.

John S. Rigden

Acknowledgements

In the Fall of 1971, discussions were held on the campus of the University of Missouri—St. Louis, which resulted in a course offering called the Physics of Music. The participants in this discussion were Professor Charles Foster of the Physics Department and Professor Neil Bjurstrom of the Fine Arts Department. Without these individuals this book would not have been written. As things turned out, it fell to me to teach the course. This book is a by-product of that responsibility.

Two students have been particularly helpful. Hewitt McElroy, a fine trombonist-turned-dentist, was an able assistant in the early days of the course. Debra Davis, a fine pianist-turned-physicist, thoughtfully read and commented upon the manuscript.

Finally, I thank Gaile Smothers who typed well and did so with a smile.

J.S.R.

St. Louis, Missouri, 1976

In memory of
James Houston Shrader

Contents

CONTENTS

Physics and the
Sound of Music

The Sound of Music

Quiet, please, listen. Regardless of where you are, what you are doing, the time of day or night, a variety of sounds are playing over your ears. Some sounds you learn to ignore totally while others bring you to instant full alert. Sound, a transom to our environment, is ubiquitous.

Among the assortment of sounds there is music. We never mistake the sounds of ocean waves, vehicular traffic, typewriters, or conversation for the sounds of music. Music can be hummed, but one cannot sing the sound of a closing door.

What is that sound we call music? Sound with a melody, a succession of tones with a coherent pattern, is musical sound. A melodic pattern can be so compelling that after hearing it only once it can be hummed from memory. You can whistle a musical sound, but you can also tap your toe to it. The succession of tones comprising a melody have a cadence, a rhythm. Melody and rhythm are characteristics of musical sound.

Yet, the question still remains: what is that sound we call music? More specifically, what is a sound we call a tone? One of our principal goals in this book is to answer these questions and to understand the answers. In

Oscilloscope trace of laboratory noises

Oscilloscope trace of a cornet playing a steady tone

the meantime, however, we might imagine monitoring some sounds with a microphone and displaying the result on a televisionlike device called an oscilloscope. By means of the curves displayed, laboratory sounds can be compared with a musical sound. The curves are clearly different. One curve has an aesthetic quality absent in the other. All musical sounds can be presented by attractive curves. When Mahler composed his *Third Symphony* we could say he was designing a curve.

Sound is a physical phenomenon. Music is an art. In the sound of music, physics and art come together and in the merger, there is unity.

1 The Physics of Music

Physics . . . music . . . a stimulating juxtaposition of words, a provoking *contrast* of images. Contrast the words that are typically employed to characterize physics vs. those used for music: quantitative vs. qualitative, objective vs. subjective, empirical vs. aesthetic, impersonal vs. personal, rational vs. emotive, head vs. heart, bloodless vs. sentimental, uncouth vs. refined, square vs. groovy. Although these words do not accurately express my thoughts, they unfortunately convey with correctness the thoughts of many people.

Music . . . physics . . . an intriguing juxtaposition of words, a provoking *synthesis* of images. Physics and music are *both* human activities with all the beauties and foibles attending any human enterprise. Physics and music *both* depend upon the creative inspiration of the human mind for their actualization. As a creative activity, physics and music share many common features. The truly significant advances in physics have had their origins more often in the realm of mind than in the domain of the laboratory. Often, ideas are born in physics long before the observations emerge that give the ideas credence. The spawning ground of both physics and music is the human mind. On the other hand, physics and music both employ observations creatively. Joseph Haydn, an eighteenth century composer, lived to see Beethoven and Mozart observe his ideas and carry on their development. Physics and music are *both* evolving endeavors: the physics of Dirac and Feynman is built upon the physics of Newton and Faraday. Likewise, the music of Copland and Shostakovich is built upon the music of Bach and Schubert.

Physics . . . music . . . delightfully different, enchantingly similar. It is my intent to bring the subjects of physics and music together. In this

3

process, I want to identify the subjective sensations of music and to understand them—to the extent possible—in terms of the physical properties of sound patterns. You shall learn some things about the physics of music. In addition, it is my hope that you will sense in part the harmony of physics.

MUSICAL SOUND— THE SYSTEMS INVOLVED

Imagine yourself listening to music. In the most general terms, this activity involves three systems. First, there is the source of the sound. Sound originates from a source. The source emits sound. Second, there is the air that propagates the sound. A medium is necessary to carry the sound from the source; it transmits the sound. Third, there is the listener that detects the sound. A detector receives the information from the source via the medium and registers that information. Origination, transmission, and detection together form a crucial part of the listening experience.

Let us briefly examine these three systems more specifically. The source of musical sound may be a piano string, a clarinet reed, or a vocal cord. By some mechanism, the source is excited into motion. We are correct in identifying a moving element as the source of musical sound. The source, however, has its own *immediate environment.* A piano string or a vocal cord by themselves are anemic sources of acoustic energy. There is a dramatic change, however, when a piano string is joined to a soundboard or a vocal cord is coupled to a mouth–nasal cavity. The immediate environment directly influences the quality and intensity of the sound that is transmitted.

The medium transmits energy from the source to the detector. Although sound can be transmitted by almost any material medium, air is of interest to us. Once again, however, the immediate environment of the air is an active participant. Boundaries reflect and absorb sound. There is a dramatic difference between *Oh Susanna* rendered in an open field and the same song sung in the shower. In both cases, air is the medium. The immediate environments, however, are very different.

The detector of interest to us is the human auditory system. The detection process begins with the eardrum that is set into motion by the air in the auditory canal. The eardrum's motion is transmitted via bone linkages to the inner ear where receptor cells initiate the neurological mechanism of perception. Here, more than before, the immediate environment of the auditory system is all important. I refer to the human brain. The information transmitted in the form of nervous impulses is assimilated by the brain. Some brains "like" chamber music, others don't.

With one notable exception, the systems involved in the act of listening to music can be understood in terms of the concepts and principles of physics. The musical instrument, the transmission of sound, room

acoustics, and auditory detection are all amenable to the techniques of physics. Furthermore, both the reproduction of musical sound and electronic music are based on physical principles. The exception comes at the neurological end of the perceptual process. Physics has nothing to say about the value judgments made by the individual listener.

Clearly, there is considerable justification in the phrase "the physics of music."

<div style="display:flex">
<div style="width:25%; text-align:right; font-weight:bold">THE WAY AHEAD— AN OVERVIEW</div>
<div style="width:75%">

Details can be interesting and even fascinating. They are important by themselves and take on added significance when they are related to form part of a larger picture. When a new subject is studied, it is difficult to maintain an overview and also to focus upon the nitty gritty. Yet this must be done. When the overview is lost, the specific details become disjointed and their significance fades or even vanishes. When the overview is lost, specific details must be remembered specifically whereas the overview, if perceived, can make sense out of a whole range of detail and do so in a straightforward fashion. If nothing else, an overview is a memory aid.

This book has a structure that in itself provides something of an overview. There are five parts, Parts II through VI, consisting of three chapters each. The first chapter of each part deals with the origination of sound, the second chapter with its transmission and the immediate environment, and the third chapter of each part is concerned with aspects of its detection and perception.

The simplest musical sound is a pure tone. The pure tone is the subject of Part II: how it originates (Chapter 2), how it is transmitted (Chapter 3), and the subjective response to it (Chapter 4). How does a speaker cone move when two pure tones, say an octave are heard coming from it? How does the medium move as the sound of these two pure tones is transmitted? These questions are answered in Chapters 5 and 6, respectively. Interesting characteristics of our auditory apparatus are encountered in Chapter 7 when we examine our ear's response to two pure tones together.

In Part IV we come to the unadorned heart of musical sound production: the stretched string and the air column. In Chapter 8, techniques learned earlier are extended to the complex vibrations of taut strings and air columns. The sound emanating from either a stretched string or an air column (Chapter 9) depends upon how the vibration is initiated and upon the immediate environment.

If we understand the oscillating string and air column, we can move on stage and consider the orchestra in its entirety. This is done in Part V. The musical instruments themselves are the subject matter of Chapter 11. The environmental influences upon the transmitted sound—room acoustics—are considered in Chapter 12. Consonance, dissonance, and

</div>
</div>

musical scales are topics of perceptual import and are discussed in Chapter 13.

More than ever before, fine music is a part of the home environment. Sound reproduction systems (Part VI) in a range of quality and a variety of configurations are available to the consumer. The components of a sound system (Chapter 14) have specifications that indicate component quality. Obviously, a sound system is to be used in a room environment. In Chapter 15 the components are considered as systems within the intended environment. Finally, buying strategies are suggested in Chapter 16. These strategies should lead to a satisfying listening experience.

Electronic music is the subject of Part VII.

Now, let's get started.

A Pure Tone—
A Capella

Just as complex molecules can be broken down into simpler units called atoms, many complex sounds can be broken down into simpler units called pure tones. A pure tone by itself sounds rather dull and uninteresting. However, if properly selected, a group of pure tones sounded together determines the quality of tone of every musical source: the velvety tone of a flute, the expressiveness of a Stradivarius, the radiance of a trumpet, the plaintively nasal character of an oboe, or the richness of an operatic tenor. Moreover, the sound of an entire symphony orchestra emanating from the speakers of our high-fidelity stereo sound systems is simply a collection of pure tones. The tinny sound of an inexpensive record player differs from the full sound of a high-fidelity system because the latter has been carefully designed to leave unaltered the many pure tones making up the total sound.

Pure tones are important, and it is with pure tones that we shall begin. To understand the physics of music, we must understand a pure tone: how it is generated (Chapter 2), how it is transmitted (Chapter 3), and how it is perceived (Chapter 4). In these chapters we shall define terminology and develop concepts about pure tones that are the basis of musical sound.

2 The Origin of Musical Sound

"God order'd motion, but ordain'd no rest." We do not know what inspired the poet Henry Vaughan to write this line, but the sources of such inspiration are manifold since the world is teeming with movement. And the world is alive with sounds, all kinds of sounds. But this is as it should be for motion and sound go together. Without motion, there would be no sound.

Music is a special kind of sound. As I write these words, I hear the music of Lara's theme from the motion picture *Doctor Zhivago* with the haunting sound of the balalaika prominent in the refrain. The balalaika is a string instrument with a triangular body and is played by plucking the strings. The strings move when they are plucked and musical sound results from this motion. As we might expect, the motion of a string is a special kind of motion.

PERIODIC MOTION

A motion that repeats itself is called a periodic motion. The moon orbiting the earth, a beating heart, a leopard pacing in its cage, the leg motions of a marching drill team, the flutter of a leaf being wafted by a gentle breeze, the whirring tail of a frightened rattlesnake, the wings of a hummingbird, the valve on a rolling bicycle tire—all are periodic motions.

(a) *(b)* *(c)* *(d)* *(e) = (a)*

Let us look a little more closely at the wings of a hummingbird. Specifically, let us focus our attention on the very tip of one wing. In (*a*) above, the wings are in their downmost position; in (*c*), they are in their upmost position, and in (*e*) they are once again in a position equivalent to (*a*). So the motion goes *a–b–c–d–a–b–c–d–a–b–c–d* . . . over and over again.

There is a very efficient way to represent the motion of the tip of the hummingbird's wing—a diagrammatic representation. In (*b*), the wings are shown at that instant in time when the wings are stretched horizontally, moving upwards. In about 0.004 sec, the wings will have moved to their uppermost position as in (*c*). Here the wing tips are a given distance above their horizontal position shown in (*b*). We shall call this distance +y. (See the figure left.) The wings will leave their uppermost position, and in about 0.008 sec will be a position (*a*)— the bottommost position. We shall denote this bottommost position as −y. Clearly, the wing tip moves back and forth from +y to −y and it is this back and forth motion that lends itself to a diagrammatic representation, such as that shown below.

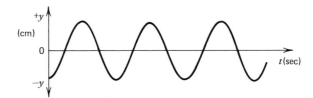

The curve in the figure shows in an approximate way the position of the wing tip (vertical axis) at every instant of time (horizontal axis). (I must say approximate because I don't know exactly how the hummingbird's wing moves.) This curve is a rather formal method of representing the motion of the wing tip. Notice that in the absence of a label there is nothing in the figure *per se* to identify it as representing the tip of a hummingbird wing. In fact, the curve might represent any number of periodic motions. It is this generality that makes the graphical technique such an efficient way of representing periodic motions. Other graphs of periodic motions are shown below. Notice that any unit of time (second, year, minute) or position (cm, meter, foot) can be used. Choice is dictated by convenience.

There are three parameters that characterize the basic features of any periodic motion. The *period, T*, is one of the parameters. The period of the earth's orbital motion is, by definition, 1 year; the period of the moon is approximately 28 days; the earth's axial rotation period is 1 day; that of the heart is about 1 sec; and the period of the hummingbird's wings is about 0.016 sec. Obviously, the period is the time required for one complete cycle of the motion. Notice that the graphs employed

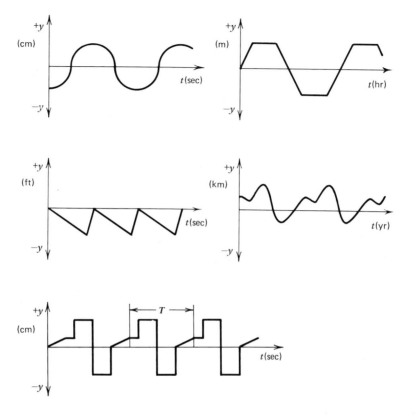

above can easily be applied to any periodic motion as long as we use the appropriate scale of distance on the vertical axis (kilometers for the moon and centimeters for the hummingbird) and the appropriate scale of time on the horizontal axis (days for the moon and fractions of seconds for the hummingbird).

The period of a cyclical motion gives us another parameter of importance. The *frequency, f,* of a periodic motion is the number of complete cycles occurring in a given increment of time. The frequency of the heart ($T \cong 1$ sec) is about 1 beat per second (we could also say 1 cycle per second); that of the moon ($T \cong 28$ days) is 1/28 days or 0.036 cycles per day; the frequency of the hummingbird's wing ($T = 0.016$ sec) is $1/0.016 = 63$ cycles per second. It should be clear that the frequency is the inverse of the period (and vice versa), that is,

$$f = \frac{1}{T} \qquad \text{or} \qquad T = \frac{1}{f}$$

The larger the period the smaller the frequency. Until recently, the frequency was most frequently expressed in cycles per second. The unit of hertz (abbreviated Hz) is now used where 1 cycle/sec = 1 Hz.

The third parameter that is important in characterizing periodic motions is the *"extent" of the motion.* The moon travels approximately 1,510,000 miles in one cycle of its motion; the earth travels much farther during one cycle of its motion about the sun, approximately 578,000,000 miles. The tip of a hummingbird's wing travels only a few centimeters during each cycle.

For many periodic motions, the greater the "extent" of the motion, the larger is the period. (Compare, for example, the periods and extents of the planets) For some periodic motions, however, the large period–large extent relationship does not hold and to this we now turn.

SIMPLE HARMONIC MOTION

Simple harmonic motion (SHM) is a very special periodic motion for which the period (or frequency) is independent of the "extent" of the motion. This special property is what makes music possible.

Later, we shall discuss a further property that makes SHM special from the point of view of anyone considering the physics of music. I anticipate this second special property to emphasize the special character of SHM and to make what may seem to be a contradictory statement: SHM is a very common motion. Special, but common. A slightly rocking boat, a vibrating guitar string, a trembling diving board, an oscillating tuning fork, a gently swinging pendulum are examples of SHM.

In order to get some idea of the attributes of SHM, let us look carefully at an arrangement that resembles the clock pendulum: a simple weight suspended from a thread. Suppose, for example, we tie a sinker to a piece of thin fish line and suspend it from a tree limb. In plumb-line fashion the sinker will jiggle for a few moments, but will soon come to rest. The sinker, hanging motionless, is in its equilibrium position; that is, if the sinker is moved slightly from this equilibrium position, it will, after some jiggling around, return to this same vertical plumb-line position. If the sinker is carefully drawn a couple of centimeters to one side and released, the sinker will oscillate back and forth through the equilibrium position. The motion takes place in a plane containing both the starting and equilibrium positions. The sinker is moving with SHM.

Equilibrium position

Start

−2 cm +2 cm

What are the attributes of SHM? As we think about this example we can list them.

1. The motion takes place symmetrically about an equilibrium position

The guitar string vibrates about the position it occupied prior to being plucked, the diving board quivers about the position it occupied prior to its being stepped upon and jumped from, and the boat rocks about the position it had before it was disturbed.

Furthermore, there is a symmetry to the motion. Considering our pendulum, we start its motion by pulling the sinker to one side, say 2 cm, and releasing it. The sinker then swings through the equilibrium position and stops 2 cm the other side of the vertical. This distance from the equilibrium position is called the *amplitude* and is symbolized by *A*. Previously we referred to the "extent" of a periodic motion, but from now on we shall refer to the amplitude of the motion. The pendulum in our example has an amplitude of 2 cm, or, if we had pulled it a distance of 1 cm to one side, the amplitude would have been 1 cm. (Looking back, can you show the amplitude on the graph of the wing tip of the hummingbird?)

2. The motion repeats itself in a definite time interval called the period

Even though the amplitude of the motion slowly decreases, the period stays the same and is thereby a definite characteristic of the motion. This is the special property of SHM.

3. The speed of the oscillator varies throughout the cycle

Specifically, the speed is zero at the extremes of the motion and is a maximum as the oscillator passes through its equilibrium position. Moreover, the larger the amplitude of the motion, the larger is that maximum speed.

There are two reasons why the simple harmonic oscillator has the three characteristics outlined above: (1) the nature of the force acting on the oscillator and (2) the inertial property of the oscillator. The force acting on the oscillator is a restoring force: always directed back to the equilibrium position. Furthermore, the force increases in direct proportion to the amplitude of the motion. The force pulling the sinker back toward the equilibrium position is twice as large when the amplitude is 2 cm than when it is 1 cm. When the sinker is in its equilibrium position, the restoring force is zero. The behavior of this force can be expressed symbolically as $F = -ky$. The minus sign indicates that the force, F, and the displacement from equilibrium, y, are oppositely directed: displacement to the right results in a restoring force to the left. The constant k indicates how difficult it is to displace the oscillator from its equilibrium

position. For example, a stiff spring has a large k while a sloppy spring has a small k.

Now we can understand something about SHM. When we pull the sinker to one side and release it, the sinker—under the influence of the restoring force—accelerates back toward the equilibrium position gaining speed all the way. It arrives at the equilibrium position and, because of its inertial property, overshoots it. As soon as it overshoots, however, the restoring force, always directed toward the equilibrium position, slows the sinker and soon brings it to rest at the opposite extreme of the motion: acceleration back toward the equilibrium position ... overshooting ... slowing down ... back to its starting point, thus completing the cycle.

It is convenient to represent SHM in the same diagrammatic fashion employed earlier. For this purpose, let's consider another example of a simple harmonic oscillator—a mass suspended from a spring. If the mass is pulled vertically downward from its equilibrium position and released, it oscillates with SHM. A graph of such a motion—with amplitude of *5 cm*—is shown below. If we push it *upward* 10 cm and release it, the graph of the ensuing motion would be different, as indicated below. Curves that correctly depict SHM are called *sine curves.*

What is the significance of SHM to music? To answer baldly, SHM is the origin of musical sounds. To be more specific, an object such as a tuning fork, oscillating with SHM gives rise to the simplest of all musical sounds—a pure tone. An object moving randomly gives rise to noise; in fact, regularity of movement is necessary if musical sound is to result. (The adjective "pure" will take on meaning in Chapter 5.)

In the case of a tuning fork, the tines of the fork oscillate with SHM when they are gently struck. The result is the tone we are used to hearing. It is clear that the tines are vibrating, as evidenced by the fuzzy edges of the tines as they whirr back and forth; moreover, if the singing tuning fork is thrust into a glass of water, its vibrating tines cause a

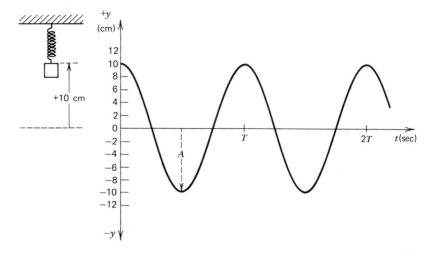

splashy display. It is possible to mount a small stylus on one tine of a fork that permits a trace of the vibrating tine to be scratched on a soot-covered surface. The path traced is a sine curve.

The tuning fork was once the basic source of pure tones. Now electronic devices called oscillators have become the principal source of pure tones. If a speaker is connected to an electronic oscillator, a pure tone is heard. Electronic synthesizers are made up of several oscillators that together can create a wide variety of sounds.

SHM is the source of pure tones and pure tones are the building blocks of music.

DAMPED AND FORCED OSCILLATIONS

Physical objects have the character of a coil spring. A steel rod can be compared to an extremely stiff spring and hence very difficult to stretch or compress. The air around us, on the other hand, is like an extremely weak spring that can be stretched or compressed with ease. The point of the analogy is this: if either the steel rod or a column of air is stretched and then released, each, like the coil spring, will oscillate. Furthermore, each will oscillate at a particular frequency characteristic of it. In fact, every object will oscillate at its own characteristic frequency, or, in most cases *several* characteristic frequencies.

Consider a block of metal suspended from a spring. If we now pull the block straight down a distance y, say 5 cm, and release it, the block will oscillate up and down with SHM. As with all SHM, the motion has a precise frequency that in this case depends on the stiffness of the spring (the constant k in the force law). A second spring, stiffer than the first, would oscillate with a greater frequency, but still a very definite frequency.

Let us consider the vibration of air. A column of air is contained in a

Amplitude = −5 cm

cylinder like that in the figure below. The cylinder is fitted with a piston that slides freely in the cylinder. Suppose the length of the air column is l and that air has been admitted to the cylinder so that the air pressure inside the column equals the pressure outside it. Now, let us "stretch" the air column by pulling the piston out a few centimeters. It may be hard to hold the piston in this stretched position because the pressure outside is greater than that inside, therefore, we release it. The piston shoots back toward the equilibrium position, but it arrives there with some speed and overshoots. As soon as it overshoots, the pressure inside the cylinder becomes greater than the pressure outside, and the piston is soon brought momentarily to rest. It starts back toward the equilibrium position . . . overshoots . . . etc. In other words, the piston, in response to the springiness of the column of air, oscillates. As in our other example, the oscillations occur with a definite frequency characteristic of the air column. If we increase the length of the column of air, the oscillation frequency decreases.

The knowledge to be gained from these considerations is that physical objects have an elastic character and when they undergo a sudden deformation—hitting a drumhead, plucking a string, blowing over a reed, bumping a spring, striking a tuning fork—they oscillate with a definite frequency characteristic of the object itself. This frequency is called the object's *natural frequency.*

Thus far we have implicitly assumed that the oscillations continue indefinitely. In fact, they don't. Present in all oscillating systems are resistive influences that deplete the energy of the oscillator with the result that the amplitude decreases and eventually the oscillator comes to rest. The very fact that we hear a tuning fork indicates that some of the vibrational energy is being transformed into acoustic energy. Thus, the air tends to dampen the oscillation of the tuning fork. A tuning fork vibrating in water would be damped to an even greater degree. The figures below compare oscillators with and without damping. The figures show that the frequencies of both the damped and undamped oscillators are equal. This equality is almost valid provided that the damping is not too large.

For some oscillators, the damping is sufficiently small so that energy

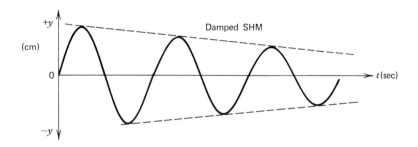

can be introduced impulsively; the resulting vibration will continue long enough to be useful in the production of a musical score. This is true of the string struck in the piano. For other oscillators the vibrations decay so quickly that energy must be supplied continuously, as is the case with the flute. Let us now examine a means of supplying energy to an oscillator.

A physical effect called *resonance* can occur when a physical object is subjected to a periodic force. Resonance is crucially important in the physics of music. We have seen that physical objects have natural vibrational frequencies. Let's now examine what happens when a periodic force is applied to an object.

A force might be a push, a pluck, or a pull, but to be periodic it must occur in a regular sequencial pattern. If, for example, a push...push...push acts on an object with a regular interval of 2 seconds, it would be a periodic force having a period of two seconds and a frequency of 0.5 Hz. When a periodic force is applied to a physical object, the object is forced to oscillate at the frequency of the applied force. Suppose, for example, the natural frequency of a pendulum is 40 cycles per minute (cpm). If a periodic force having a frequency of 30 cpm is applied to the pendulum it will, after an initial period of instability, oscillate with a frequency of 30 cpm. When the applied force pulls to the left, the pendulum will move to the left. The amplitude of the pendulum will be small. If the frequency of the periodic force is now changed to 50 cpm, an interesting change occurs. Again, after the initial period of instability, the pendulum will oscillate with small amplitude at

a frequency of 50 cpm; however, when the applied force is pulling to the left, the pendulum will be swinging to the right! When the frequency of the driving force is less than the natural frequency of the pendulum, it swings *in phase* with the driving force. On the other hand, when the frequency of the driving force is greater than the natural frequency of the pendulum, it swings *out of phase*. This phase relationship will be important later.

When the frequency of the applied force differs from the natural frequency of the driven object, the force and object actually work against each other with the result that the amplitude of the forced oscillation is small. On the other hand, if the frequency of the applied periodic force is equal to the natural frequency of the object, large-amplitude oscillations result.

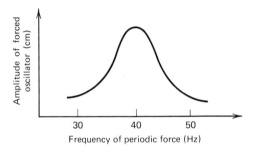

Response Curve. When the frequency of the applied force (horizontal axis) is near the natural frequency of the pendulum (40 Hz), the resulting amplitude of the pendulum (vertical axis) is large.

Children are unknowingly familiar with this phenomenon. Suppose for a moment a child is sitting in a swing. If the child is pulled back from the vertical equilibrium position and released, back and forth oscillation will occur. If a second child (our periodic force) pushes the swing, the frequency of the pushes is of necessity equal to the natural frequency of the swing and the result is an oscillation whose amplitude increases with each push. Imagine the outcome if pushes could be randomly applied to the swing.

The creation of large-amplitude vibrations in an object by an applied periodic force whose frequency equals the natural frequency of the object is called resonance (the force itself may be very small). A tuning fork alone sounds rather feeble, but when placed on a hollow wooden box designed so that the air inside has a natural oscillation frequency equal to the tuning fork, a much louder sound is heard. Why? The tuning fork is now the periodic force and it drives the air in the box, which quickly responds with large-amplitude oscillations. The box, in this case, is often called a resonator and amplifies the sound of the tuning fork.

The body of a violin is also a resonator. When a violin string is bowed, it oscillates. The string oscillations set the bridge into oscillation and the bridge drives the body of the violin, which in turn drives the air enclosed by its body. Both the enclosed air and the body of the fiddle have frequencies at which they tend to vibrate naturally (there are many natural frequencies for both the air and the fiddle body) so that the enclosed air and fiddle body can resonate if the bridge drives them at the correct frequency. A violin string stretched between two sides of a window frame would have a meager sound when bowed, but when placed on a properly designed resonating box gives a familiar result.

Note that resonance can have disagreeable effects. For example, a musical instrument may set furniture, knickknacks, or windows in a room into vibration. (Here the frequency of the sound is the same as the natural frequency of a knickknack and the sound drives the object.) Later we shall discuss the piano; in the meantime, however, think about the piano's resonator—its soundboard—and some of the demands it must satisfy.

For Further Thought

1. Graphically represent the motions of:
 (a) A moving teeter-totter
 (b) A vibrating tuning fork
 Carefully label each axis and identify the units being employed.

2. The moon's motion is periodic. Think of a way to represent the motion of the moon graphically. Identify the units being employed and label the axes.

3. A simple harmonic oscillator (SHO) has a period of 0.002273 second. What is the frequency of the oscillator?

4. A SHO has the following characteristics:

$$\text{Amplitude} = 2.0 \, \text{cm}$$
$$\text{Frequency} = 150 \, \text{Hz}$$

 Make a graph of this motion plotting y against t.

5. The force law $F = -ky$ means that the larger the displacement from equilibrium (y), the larger is the restoring force (F). Force is expressed in units of *newtons*. (Roughly speaking, 1 newton is about the force required to balance the gravitational force acting on a large glass of water.) If a spring has a stiffness constant of 71 newtons/m, what is the force required to hold the spring in a stretched position 0.075 m beyond its equilibrium position?

6. A SHO has a natural frequency of 200 Hz. It is slightly damped. A periodic force is applied to the oscillator. Describe the motion of the oscillator when the applied force has a period of:
 (a) 0.00556 sec
 (b) 0.00455 sec
 (c) 0.00500 sec

7. The shape of a response curve is dependent upon the amount of damping present in the oscillator. The smaller the damping, the more sharply peaked the response curve. In other words, the smaller the damping, the larger is the response of the oscillator when it is driven at the resonant frequency.
 (a) Which would be more responsive at frequencies off resonance, an oscillator with large damping or one with small damping?
 (b) Can you anticipate the acoustical significance of the answer you gave in (a)? We shall understand the significance later.

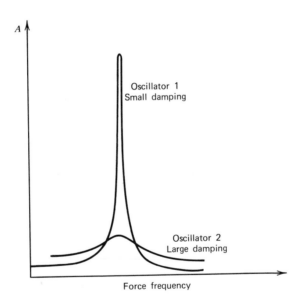

3 The Transmission of Musical Sound

Sound originates with motion and musical sound originates with simple harmonic motion. If, however, we had a tuning fork vibrating on the moon, there would be no sound. Sound, as the word is commonly understood, results from a wave that is sent coursing through the medium in which the tuning fork is vibrating. Ordinarily, of course, the medium is air.

Sound is transmitted by a medium. While air is commonly the medium, any material will suffice. Water transmits sound. Steel and bone transmit sound. Gases, liquids, and solids transmit sound. In all cases a disturbance in the form of an elastic compression or rarefaction is propagated through a medium. Generally speaking, the more rigid the medium, the more quickly the disturbance travels. And the disturbance travels until it is dissipated into heat within the medium or until it encounters a boundary—a wall, an edge—at which it is partially reflected and partially transmitted into the domain beyond the boundary.

PROPAGATING DISTURBANCES IN SPRINGY MATTER

Because of the springiness of matter, a disturbance produced in one part of an object is transmitted to other parts. Physicists refer to the disturbance as a wave and to the substance in which the disturbance is moving as the medium. The medium is not a passive element in this process; on the contrary it is the medium that propagates the distur-

21

bance. A wave, then, is a disturbance propagated by a material medium.*

What is meant by a disturbance? Stand a long line of dominos on end facing each other and gently push the end domino into the next. What happens? The end domino pushes over its neighbor, which in turn pushes over the next domino and so on until "the disturbance" has passed through the line of dominos. A better answer to the question is afforded by the experience of tossing a pebble into a quiet pond. What happens? The pebble sets up a disturbance in the medium (water) in the form of ripples that spread across the pond as circles with increasing radii. Or consider a rope stretched rather taut. A disturbance can be initiated by quickly snapping one end of the rope up and down. What happens? The disturbance, in the form of a hump in the rope, propagates along the length of the rope.

In order to better understand the mechanism that propagates a wave in a medium, let us consider a model. To simplify our model we shall make it one dimensional—a line of beads connected by small springs. In our model, illustrated below, each bead has one spring pulling it to the right and another pulling it to the left. As it stands, however, each bead is in equilibrium; that is, the pull to the left balances the pull to the right.

Now let us disturb the system by suddenly pulling the piston—to which the left end of the line of beads is fastened—to the left. What happens? Spring 1 is stretched, which increases the force pulling bead *a* to the left and, in response, bead *a* moves in that direction. Spring 2 is then stretched and bead *b* is forced to move to the left thereby stretching spring 3. And so it goes. Each bead responds to the spring forces acting upon it and in so doing a wave is propagated along the line of beads.

Let us notice one thing: the wave propagates from the site of the disturbance to the other end of the medium; however, the motion of the medium itself is very limited. This is the very essence of wave motion. It involves the propagation of a disturbance, but *not* the transport of the medium itself. In other words, bead *a* never ends up at the right-hand end of the line.

*Light is a special kind of wave that requires no medium. Sunlight successfully gets from the sun to the earth through the vacuum of space. We shall not consider light waves any further.

Two kinds of waves are important in the production of musical sound. We can illustrate both of these with the bead model. In fact, we have already illustrated one of them. When the piston was pulled to the left, a wave was initiated and propagated along the line of beads. During this process, however, the beads themselves moved in a direction parallel to the direction of the wave motion. The wave moved to the right and the beads moved along a left-to-right line. Such a wave, when elements of the medium move *parallel* to the direction of propagation, is called a *longitudinal wave.*

The second kind of wave is called a *transverse wave.* To observe it we must initiate the disturbance differently. Let us exert a sudden vertical upward force on bead *a* so that, for a brief moment, it looks like the figure below.

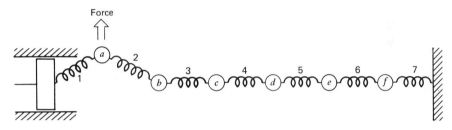

With the disturbance initiated in the above fashion, spring 2 exerts an upward force on bead *b* and, in response to this force, bead *b* moves upward. Then an upward force is exerted by spring 3 on bead *c*, etc. At some time after the disturbance is initiated, the line of beads looks like this:

It should be apparent that in a transverse wave, elements of the medium move in directions *perpendicular* to that of the wave motion.

When a cello string is bowed, a transverse wave propagates along the string. The vibrations of the string are transmitted to the body of the cello and the neighboring air is thereby disturbed. The result is a sound wave.

SOUND WAVES Sound waves are longitudinal waves.

The source of musical sound is a vibrating object: a vocal cord, an air column, or a string. Whatever the vibrating element, the contiguous air

is periodically compressed and rarefied and a disturbance is transmitted from the source in all directions as a longitudinal wave. As the sound wave moves through the air successive regions of the medium are compressed and rarefied with the result that the pressure at a point in the path of a sound wave varies periodically. It is this periodic pressure variation that ultimately enters our auditory canal and initiates the chain of events leading to perception.

Mechanism of a Wave

In a rather straightforward way we can understand how a sound wave works. Consider a long tube that is open at one end and contains a piston at the other end. The air in the tube is at atmospheric pressure. Our system is pictured below. Suppose the piston is quickly thrust into the

tube, say 1 cm. As a consequence of the piston's sudden motion two things happen: first, the molecules of air pile up in front of the piston with the result that the pressure immediately in front of the piston is higher than the atmospheric pressure of the rest of the air. In other words, a small portion of the air is compressed. But a second thing happens: the moving piston collides with the molecules of the air and imparts to them a large velocity directed away from the piston. This high-velocity group of molecules collides with other molecules and thereby the high velocity is transferred to them. As a consequence of the high-velocity transfer, a well-defined compression wave propagates down the tube. This disturbance is a longitudinal wave pulse—but in this case a very unmusical sound wave.

Molecules pile up in front of the piston as it is thrust into the cylinder

Compression propagates through the gas

An analogy might be useful. Imagine a dance floor crowded with waltzing couples relaxed by the mood of the background music. On the fringe of the dancing area a couple stumbles into other dancers who in turn push into still other dancers, etc. In this fashion, a longitudinal wave pulse propagates across the dance floor. Like the molecules of the

air, the dancers themselves are not transported to any extent; rather, it is a disturbance that is transported through the medium.

A musical sound wave is generated by a periodic disturbance. For example, imagine the piston to be driven by the simple harmonic motion of a tuning fork tine as illustrated below. The piston is alternately

pushed into and pulled out of the tube, and a succession of compressions and rarefactions propagates down the tube. As before, a compression results when the piston moves into the tube and through velocity transfer, it travels down the tube. When the piston is pulled out of the tube, a low-pressure region called a rarefaction is created in front of the

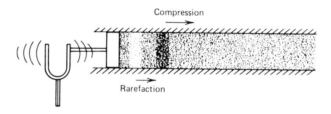

piston. Molecules move into the low-pressure region, thereby creating a new low-pressure region a little further down the tube. In this manner the rarefaction propagates down the tube. After several cycles of the piston, compressions and rarefactions are moving through the tube in a fashion illustrated below.

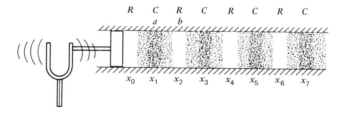

We can represent a periodic sound wave more efficiently by graphical techniques. Before the piston begins its back and forth motion, the air in the tube is at atmospheric pressure. The piston's periodic motion creates a succession of compressions (pressure above atmospheric pressure) and rarefactions (pressure below atmospheric pressure) that propagate through the medium. Thus the sound wave can be pictured as pressure variations within the medium. In a pure-tone sound wave the pressure

variations above and below atmospheric pressure follow a sine curve—the same kind of curve describing the amplitude variations of a simple harmonic oscillator. A graphical representation of it is shown below. The graph represents the pressure variations at a particular instant of time:

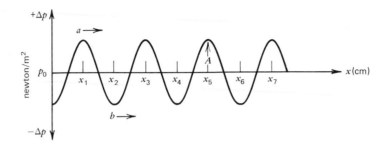

when compression a is located at position x_1 (see previous figure) rarefaction b is located at position x_2, etc. At a later time, compression a, rarefaction b, etc. will have moved along the tube to new positions. The amplitude, A, of the sound wave is the magnitude of the pressure variation above and below the ambient atmospheric pressure.

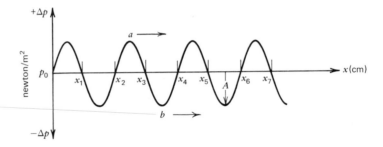

Parameters of a Wave

Several parameters characterize a sound wave (or any other wave). The *amplitude* is one such parameter. For a sound wave it is a pressure amplitude; that is, the greater the pressure variations above and below atmospheric pressure, the greater the amplitude of the wave. *Wavelength* is another parameter. Symbolized by λ (the Greek letter lambda), wavelength is the distance between the centers of two adjacent compressions or, if we ignore the maximum and minimum pressure points, wavelength is the distance between every other spatial point having the same pressure amplitude.

The *frequency* of a wave is a parameter that can be understood in several ways. For example, as a sound wave propagates along the tube, the frequency is simply the number of compressions (or rarefactions) that pass a particular point in the tube every second. If the sound is that

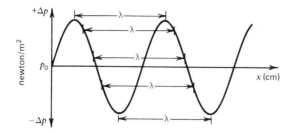

of middle C, 262 compressions pass a point in the tube each second. On the other hand, the frequency can be understood by the motion of the medium. When a periodic disturbance propagates through air, the molecules of the medium move back and forth (parallel to the direction of the propagating wave) periodically. The frequency of their periodic motion is also the frequency of the wave. Finally and most significantly, the frequency of the source is the same as the frequency of the resulting wave. Therefore, if we know the frequency of the source, we know the frequency of the wave. As mentioned before, the *period*, in this case of a wave, is the inverse of the frequency.

The last sound-wave parameter is the *speed* with which it propagates. In a simple way we can arrive at a very general expression for the speed of any periodic disturbance be they rope waves, water waves, or sound waves. This expression relates the speed of the advancing wave to its wavelength and frequency. Since a wave advances a distance of one wavelength in a time interval of one period, it follows that the speed of the wave is given by

$$\text{speed} = v = \frac{\text{distance advanced}}{\text{time it takes to advance}} = \frac{\lambda}{T}$$

But since $T = 1/f$, we can write

$$v = \frac{\lambda}{T} = f\lambda$$

How can the above expression be used? The frequency of a wave is determined by the frequency of the source; therefore, f is generally known. How about λ? We need something else. The speed of a sound wave traveling through air depends on the temperature of the air. For this reason, among others, performers in a symphony orchestra warm up their instruments before they tune to the principal oboe. The dependence of sound speed on temperature is expressed most easily by the absolute temperature scale. Thus, the speed of sound waves in air (in m/sec) can be written

$$v = 20.1\sqrt{T_A}$$

where T_A is the absolute temperature and the constant 20.1 is determined from basic properties of air. [The absolute temperature can be obtained from the Celsius temperature by the relation $T_A = T_C + 273°$ or from the Fahrenheit temperature by the relation $T_A = 273° + \frac{5}{9}(T_F - 32)$.] At room temperature ($T_F = 70°$ or $T_C = 21°$) the speed of sound is 345 m/sec (or 1131 ft/sec) while at the "freezing temperature" ($T_F = 32°$ or $T_C = 0°$) it is 332.5 m/sec (= 1089 ft/sec).

We now have two expressions for the speed of sound.

$$v = f\lambda \qquad \text{and} \qquad v = 20.1\sqrt{T_A}$$

whose right-hand sides can be combined to give a third relation:

$$f\lambda = 20.1\sqrt{T_A}$$

or

$$\lambda = \frac{20.1}{f}\sqrt{T_A} \qquad (\lambda \text{ is in meters})$$

The lowest C on the piano, C_1, has a frequency of 32.7 Hz. The above expression tells us that, at 21°C = 294° absolute, the wavelength is 10.5 m (or 34.5 ft). Middle C, C_4, has a frequency 261.6 Hz and a wavelength—at 21°C—of 1.32 m (or 4 ft, $3\frac{1}{2}$ in.). The highest C on the piano, C_8, has a frequency of 4186 Hz and a wavelength at room temperature of 0.082 m (= $3\frac{1}{8}$ in.).

Behavioral Characteristics of Waves

Waves behave in characteristic ways: they can be *reflected* and refracted; they can undergo *diffraction*; and, if two or more waves are traversing the same medium, they can *interfere* with each other. Since refraction does not play an important role in the physics of musical sound (except under certain conditions when music performances take place out of doors), we shall not discuss it further. Interference is important, but we shall investigate this topic in a later chapter. That leaves reflection and diffraction to be examined in this section.

When a sound wave encounters a boundary such as a wall, ceiling, or floor, it is partially reflected and partially absorbed or transmitted. The relative proportions reflected and absorbed depend on the nature of the material forming the boundary. A smooth hard substance reflects most of the incident sound wave while a porous substance like acoustical tile absorbs a large fraction of the acoustical energy.

In the reflection process the reflection angle, $<r$, is equal to the incident angle, $<i$. This is illustrated below where successive compressions are shown at right angles to the direction of the traveling wave.

Waves tend to bend around obstacles and corners. This behavior is called diffraction. The extent of the bending depends on the wavelength of the waves and the dimension of the obstacle. In particular, the degree

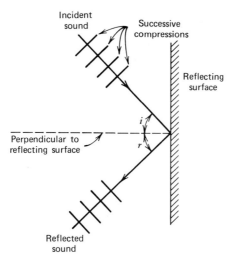

of bending depends on the ratio of the wavelength to the obstacle dimension, that is,

$$\frac{\text{wavelength}}{\text{obstacle dimension}}$$

The larger this ratio, the greater the extent of the bending. Sound waves have wavelengths on the order of a meter. Light waves, on the other hand, have wavelengths on the order of 0.00005 cm. Will sound waves and light waves diffract around a tree? A tree trunk is approximately $\frac{1}{2}$ meter in diameter. For a sound wave this ratio is much larger than it is for a light wave. Result? We can hear around trees but we cannot see around trees. An open door can be considered as a special kind of obstacle. Sound entering an open door spreads through the room (bending around the corner) as illustrated below. Yes, you can *hear*

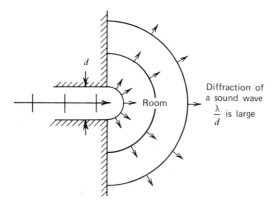

around corners. Light waves, however, do not undergo any appreciable diffraction when passing through an open door. No, you *cannot* see around corners.

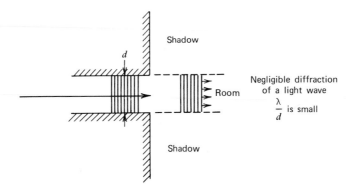

Diffraction plays a significant role in the design of speaker systems. Low-frequency sound has a large wavelength, high-frequency sound, a small wavelength. Thus, as sound emanates from a speaker, the large wavelength (low tones) sounds spread out, because of diffraction, to a greater degree than the small wavelength (high tones) sounds. In other words, the high tones tend to be more concentrated along the axis of the speaker. To compensate for this diffraction effect, speaker systems have several small speakers designed for high-frequency sound (these speakers are called tweeters) arranged in a fanlike configuration so that the high tones get spread out along with the low-frequency sound emanating from one large speaker (called a woofer).

INTENSITY OF SOUND WAVES

Musical sound originates with a vibrating source that disturbs the air in a periodic fashion and sends a sound wave traveling through it. Put another way, the vibrating source transfers energy to the medium (air) and this energy propagates through the air in the form of a sound wave. Or we could say that the sound source radiates acoustical energy.

Musical instruments radiate acoustical energy. The *rate* at which an instrument radiates acoustical energy is that instrument's power output. Power is important in the physics of music. Our ears, for example, are sensitive to the rate at which energy arrives at the eardrum; that is, they are sensitive to acoustical power. The unit of power is the watt. (A 150-watt light bulb utilizes energy at twice the rate of a 75-watt bulb.)

A source of sound emits sound waves (radiates acoustical energy) in all directions. The sensation of loudness is determined by acoustic power divided by area, that is, intensity. The *intensity* of a sound wave is the power (energy/sec) transmitted through an area of 1 m^2 oriented perpendicular

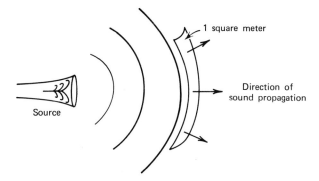

to the propagation direction of the wave (see illustration below). We shall symbolize the intensity as I expressed in watt/m². The greater the intensity, the greater is the perceived loudness.

Almost everyone knows that as you move away from a sound source, the perceived loudness decreases. This observation can be directly understood according to our definition of intensity. Consider the figure below. The sound waves propagate through the 1 m² area situated 10 m from the source. The intensity of the sound waves is, by definition, the power transmitted through that area. But look at the figure: the power concentrated over 1 m² at a distance of 10 m is spread over a larger area at the distance of 20 m. This means the intensity has decreased. Specifically, the area at 20 m is four times the area at 10 m so that the intensity at 20 m is one fourth the intensity at 10 m. Doubling the

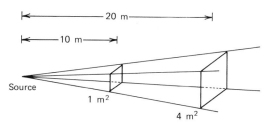

distance from the source *decreases* the intensity by a factor of 4; tripling the distance *decreases* the intensity by a factor of 9. These results can be generalized by saying that the intensity is inversely proportional to the square of the distance from the source, that is,

$$I \propto \frac{1}{r^2}$$

So the decrease in loudness as one moves away from a sound source can be understood in terms of the $1/r^2$ character of the intensity.

Now let's anticipate some of the perceptual characteristics that we will discuss more fully in the next chapter. If we listen to a pure tone of frequency 1000 Hz (two octaves above middle C), to what range of intensities do our ears respond? A 1000 Hz tone with an intensity $I = 1$ watt/m^2 strikes us as extremely loud. In fact, it hurts our ears. This defines the *threshold of feeling*, the high intensity extreme of our ear's normal operating range. (At larger intensities our ears can be damaged.) The other extreme is called the *threshold of hearing*. We can just hear a 1000 Hz tone with an intensity of $I = 10^{-12}$ watt/m^2. Think for a moment about this: the intensity at the threshold of feeling is one trillion (10^{12}) times greater than the intensity at the threshold of hearing. Some instruments, these ears of ours! Some intermediate intensities along with the appropriate musical notation are shown in Table 1.

Table 1

Intensity (watt/m^2)	Loudness
1	Threshold of Feeling
10^{-3}	*fff*
10^{-4}	*ff*
10^{-5}	*f*
10^{-6}	*mf*
10^{-7}	*p*
10^{-8}	*pp*
10^{-9}	*ppp*
10^{-12}	Threshold of hearing

fff	(No Italian word)	Extremely loud(?)
ff	fortissimo	Very loud
f	forte	Loud
mf	mezzo forte	Moderately loud
p	piano	Soft
pp	pianissimo	Very soft
ppp	(No Italian word)	Extremely soft(?)

Because the ear is sensitive to such an enormous range of intensities, the unit watt/m^2 necessitates the use of unwieldy numbers. Also, we shall normally be more interested in comparing two intensities through their ratio. (If, for example, the ratio of two intensities is 2, one intensity is twice that of the other.) For these reasons it will be convenient to introduce a new measure for the intensity of sound.

This new measure is called the *sound-intensity level, SL*, with units of decibels (db). Since the sound-intensity level is a logarithmic function, let us take a brief excursion from our main pursuit and learn something about logarithms.

Logarithms The logarithm of a number is the exponent to which 10 must be raised to yield that number. For example, 2 is the exponent to which 10 must be raised to yield 100 ($10^2 = 100$); therefore, the logarithm of 100 is 2. This is written

$$\log 100 = 2$$

Here are similar examples:

$$1000 = 10^3 \rightarrow \log 1000 = 3$$
$$100 = 10^2 \rightarrow \log 100 = 2$$
$$10 = 10^1 \rightarrow \log 10 = 1$$
$$1 = 10^0 \rightarrow \log 1 = 0$$
$$0.1 = 10^{-1} \rightarrow \log 0.1 = -1$$
$$0.01 = 10^{-2} \rightarrow \log 0.01 = -2$$

Since the logarithms of numbers between 1 and 10 are a little less obvious; we have given these values in Table 2.

Table 2

Number	Log of Number
1	0
2	0.301
3	0.477
4	0.602
5	0.699
6	0.778
7	0.845
8	0.903
9	0.954
10	1.000

From the above table we see, for example, that, $\log 7 = 0.845$. With this brief survey we can continue with our subject.

The sound-intensity level is defined as follows:

$$SL \text{ (in db)} = 10 \log \frac{I}{I_0}$$

where I_0 is the intensity at the threshold of hearing. At 1000 Hz, I_0 is 10^{-12} watt/m^2. (See Table 1.) I_0 plays the role of a reference intensity; that is, larger intensities are compared to I_0 through the ratio I/I_0. Now let's determine some sound-intensity levels.

1. *SL* of *I* = 1 watt/m² (threshold of feeling)

$$SL = 10 \log \frac{I}{I_0} = 10 \log \frac{1 \text{ watt/m}^2}{10^{-12} \text{ watt/m}^2}$$

$$= 10 \log \frac{1 \times 10^{12}}{10^{-12} \times 10^{12}} = 10 \log \frac{10^{12}}{1}$$

$$= 10(\log 10^{12}) = 10(12) = 120 \text{ db}$$

$$(\log 10^{12} = 12)$$

2. *SL* of *I* = 10⁻¹² watt/m² (threshold of hearing)

$$SL = 10 \log \frac{I}{I_0} = 10 \log \frac{10^{-12} \text{ watt/m}^2}{10^{-12} \text{ watt/m}^2}$$

$$= 10(\log 1) = 10(0) = 0 \text{ db}$$

Note. The audio range in terms of intensity increased from 10^{-12} watt/m² to 1 watt/m². The scale of sound-intensity levels is compressed between 0 and 120 db, which means much more convenient numbers to work with!

3. *SL* of *I* = 10⁻³ watt/m² (*fff*).

$$SL = 10 \log \frac{10^{-3} \text{ watt/m}^2}{10^{-12} \text{ watt/m}^2} = 10(\log 10^9)$$

$$= 10(9) = 90 \text{ db}$$

(We can say that *fff* is 90 db above threshold.)

4. *SL* of *I* = 10⁻⁹ watt/m² (*ppp*).

$$SL = 10 \log \frac{10^{-9} \text{ watt/m}^2}{10^{-12} \text{ watt/m}^2} = 10(\log 10^3)$$

$$= 10(3) = 30 \text{ db}$$

In the same fashion we could continue to determine SLs. The results are given in Table 3. Try to verify the numbers there. Also, compare Tables 1 and 3.

Table 3

SL (in db)	Loudness
120	Threshold of feeling
90	*fff*
80	*ff*
70	*f*
60	*mf*
50	*p*
40	*pp*
30	*ppp*
0	Threshold of hearing

Let us ask another question: what is the change in *SL* when *I* is doubled? In other words, what is the change in *SL* when two trumpets play the same note, each with the same intensity as played previously by only one trumpet? Here one trumpet plays with intensity *I* and two trumpets result in an intensity 2*I*. Let us adopt one trumpet's intensity as our reference. The change in *SL* when a second trumpet is added is given by

$$SL = 10 \log \frac{I}{I_{ref}} = 10 \log \frac{2I}{I}$$
$$= 10(\log 2) = 10(0.3) = 3 \text{ db}$$

Thus we see that a doubling of the intensity increases the sound-intensity level by 3 db.

For Further Thought

1. Two waves are initiated: one wave in a long steel rod and the other in a stretched wire. The wave in the steel rod is initiated by striking the end of the rod with a hammer. In the stretched wire the wave is initiated by striking the taut wire 20 cm from one end. Would you expect these waves to be longitudinal or transverse waves? Explain.

2. A sound wave was represented graphically on page 26 by plotting Δp vs. *x*. Show how a Δp vs. *t* graph could be used to represent a sound wave. Draw such a graph and carefully explain its meaning.

3. A sound wave has a frequency of 131 Hz. If the temperature is 22°C, what is the wavelength of the wave?

4. A sound wave in a medium (not air) has a wavelength of 50 meters and a frequency of 220 Hz. What is the speed of sound in this medium?

5. A sound source far removed from reflecting surfaces has an intensity measured to be 0.5 watt/m² at a distance 20 m from the source. What would be the intensity of this sound if it were measured at a distance of:

 (a) 10 meters from the source?
 (b) 30 meters from the source?

6. A sound has an intensity of 10^{-5} watt/m². What is the *SL* of this sound?

7. A trombone plays a note with a sound-intensity level of 85 db. If two more trombones join the first instrument and each plays with a *SL* of 85 db, what is the sound-intensity level of all three instruments combined?

8. The intensity of a wave increases when the amplitude increases. The relationship between intensity and amplitude, however, is not a linear one; that is,

Intensity is *not* directly proportional to amplitude.

Rather, if we identify an intensity I_1 with a pressure amplitude Δp_1, then we can write:

An amplitude $\Delta p_1 \rightarrow$ gives rise to an intensity I_1

An amplitude $2\Delta p_1 \rightarrow$ gives rise to an intensity $4I_1$

An amplitude $3\Delta p_1 \rightarrow$ gives rise to an intensity $9I_1$

An amplitude $n\,\Delta p_1 \rightarrow$ gives rise to an intensity $n^2 I_1$

In other words,

$$\text{intensity} \propto (\text{amplitude})^2$$

In comparing the intensities of two waves with their respective amplitudes we can write

$$\frac{I_1}{I_2} = \frac{(\Delta p_1)^2}{(\Delta p_2)^2}$$

Let us assume that an amplitude $\Delta p_1 = 1$ (in arbitrary units) gives rise to an intensity $I_1 = 1$ (in arbitrary units). By how much must the amplitude increase if the intensity is to be tripled?

4 The Perception of a Pure Tone

The problem of perception brings together the subjects of physics, physiology, and psychology. It is a complicated problem and remains at the frontier of current research. For example, one of the most obvious psychological attributes of musical sound is pitch. What is it about sound that determines its pitch? Clearly, the pitch of a sound wave is closely related to its frequency; however, Wightman and Green have pointed out that the exact nature of the relation between pitch and frequency remains a mystery.* Hence, there remains no completely satisfactory answer to the "simple" question posed above.

We shall limit our treatment of perception in this and succeeding chapters to the nature of the stimulus received and to the way in which this stimulus is coupled to the auditory system. In this way, we shall be able to identify the physical characteristics of the stimulus that are responsible for the perceived effect. We shall make no attempt to analyze the physiological, neurological, or psychological elements that enter into the perception process.*

THE ROLE OF FREQUENCY: PITCH

The human auditory system is sensitive to a limited range of frequencies. The typical young adult can hear sounds in the range 15 to 20,000 Hz. This wide frequency band—a little over 10 octaves—representing the region of continuous audibility is called the sonic range. Sound waves with frequencies less than 15 Hz and greater than

*See reference at the end of this chapter.

20,000 Hz are called infrasonic and ultrasonic, respectively, and are inaudible.

Infrasonic		Sonic Region	Ultrasonic	
low	15 Hz		20,000 Hz	high

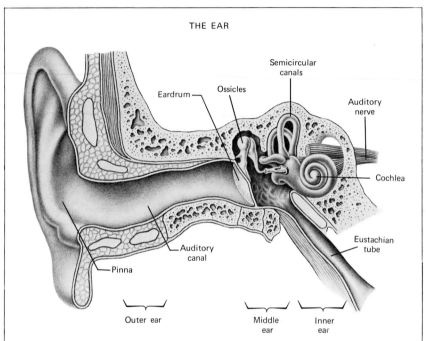

THE EAR

Semicircular canals

Eardrum Ossicles Auditory nerve

Cochlea

Eustachian tube

Auditory canal

Pinna

Outer ear Middle ear Inner ear

(From Fig. 71 "Auditory Mechanism" in *Anatomy for Speech and Hearing*, 2nd Ed. by John M. Palmer. Copyright © 1972 by John M. Palmer. By permission of Harper & Row, Publishers.)

The anatomical structure of the ear can be conveniently divided into three parts: the outer ear, the middle ear, and the inner ear. The outer ear consists of the external portion called the pinna, the auditory canal that is approximately 3 cm long, and the membrane at the inner end of the auditory canal called the eardrum. The middle ear begins just inside the eardrum and consists of a chain of three bones called the ossicles: the hammer, the anvil, and the stirrup. Opening into the middle ear from the throat is the Eustachian tube that permits equal pressures to be maintained on each side of the eardrum.

The stirrup links the anvil (on the middle ear side) to the round window, which is the beginning of the inner ear. The inner ear is a liquid-filled, coiled up cavity called the cochlea. If the cochlea were uncoiled, it would have a length of about 3.5 cm. Dividing the cochlea along its length is the basilar membrane. Hairlike cells line the basilar membrane and these hair cells are "activated" in the perception process.

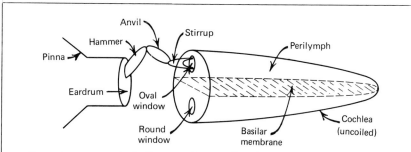

Sound waves start the eardrum vibrating. These vibrations are transmitted to the round window by the bone chain of the middle ear. The vibrating round window systematically disturbs the fluid (called the perilymph) in the inner ear and these disturbances propagate through the cochlea. The basilar membrane is set into motion by the fluid and the hair cells pick up this motion. The motions of the hair cells initiate signals in the nervous system that lead to perception.

If the acoustic signal is a pure tone, only a few hair cells on a particular portion of the basilar membrane are activated. In fact, as the frequency changes, the portion of the basilar membrane that is activated also changes. Low frequencies activate hair cells on the far end of the basilar membrane while high frequencies activate hair cells on the end nearer the round window.

Musical instruments generate sounds having frequencies ranging from approximately 20 to 5000 Hz. The lowest note on a piano, A_0, has a frequency of 27.5 Hz while the highest note, C_8 (C_8 is seven octaves above the lowest C on the piano, which is C_1), has a frequency of 4186 Hz. Orchestral tones encompass the lowest note of the contra bassoon (A_0, 27.5 Hz) and B_7 of the piccolo (B^\flat, 3729.3 Hz). All musical instruments generate harmonics (which we shall discuss thoroughly later) that are crucial to the character of musical sound. The frequencies of harmonics are musically significant up to at least 12,000 Hz.

The eardrum, or tympanic membrane, is a thin membrane between the outer and middle ear. (See the previous illustration.) Air, at atmospheric pressure, is in contact with both sides of the eardrum. (The middle ear is in contact with the outside air via the Eustachian tube.) When a sound wave sets up a succession of compressions and rarefactions in the auditory canal of the outer ear, a pressure differential is established across the eardrum and it is set into vibratory in-and-out motion: a compression forces the eardrum *in* and a rarefaction forces the eardrum *out*. The vibrations of the eardrum are transmitted by the bone structure of the middle ear to the inner ear where the stimuli are processed mechanically and neurologically. When these vibrations have a frequency in the range $15 \leq f \leq 20,000$, nerve impulses are initiated in

the inner ear that are interpreted by the brain as sound. If the pressure in the auditory canal varies in a sinusoidal fashion and if the amplitude of the pressure variation is small, the eardrum will oscillate with SHM—constant frequency and amplitude—and a pure tone will be perceived. The frequency of the oscillation we perceive as pitch: the larger the frequency, the higher the pitch. A pressure oscillation with a frequency of 440 Hz has the same pitch as A_4 (A above middle C) on the piano.

THE ROLE OF AMPLITUDE: LOUDNESS

The greater the magnitude of the pressure variations in our auditory canal—that is, the greater the amplitude of the wave—the louder is the perceived sound.

The intensity, I, and the sound-intensity level, SL, discussed in the previous chapter, are both purely physical quantities whose magnitudes are a measure of the acoustic energy propagated by a sound wave. Both I and SL are purely physical quantities because they can both be measured with no reference to the ear. Actually, loudness is very subjective and ear dependent. The problem is simply this: the ear does not respond equally to all frequencies. In other words, two tones with the *same SL* but with different frequencies are judged to have *different* loudnesses. It will therefore be useful to have, in addition to the objective measures I and SL, a subjective-based system of measure.

A subjective measure of loudness can be established by determining the SL of different frequencies judged by a number of individuals to be equally loud. Such experiments proceed as follows. First, our ears are very sensitive at frequencies around 1000 Hz. Therefore, 1000 Hz is taken as a standard reference frequency. Second, listen to a 1000 Hz tone with a $SL = 40$ db (pianissimo). Third, listen to a second tone, say a tone of frequency 100 Hz, and adjust its loudness until it is judged to have the same loudness as the 40 db, 1000 Hz tone. Measure the SL of the 100 Hz tone. *Result*: $SL = 62$ db (see point A in graph below). Fourth, repeat step three for another frequency, say $f = 30$ Hz. *Result*: $SL = 80$ db (point B). Fifth, repeat step three for many other frequencies. Sixth, on a logarithmic graph of SL vs. f, plot these points obtained in steps three, four, and five and draw a line through these points.

The curve resulting from step six above is clearly an *equal loudness curve*. Such a curve is shown in Figure 1. Now we shall define a new term: *loudness level* symbolized LL and having the unit of *phon*. All tones on the curve in Figure 1 have the same LL, namely, 40 phons. In other words, the LL (in phons) of a tone of frequency f is equal to the SL (in db) of a 1000 Hz tone judged to be equally loud. We can summarize our results in the following manner:

For a tone of $f = 1000$ Hz (remember, 1000 Hz is our reference frequency):

$$LL \text{ in phons} = SL \text{ in db}$$

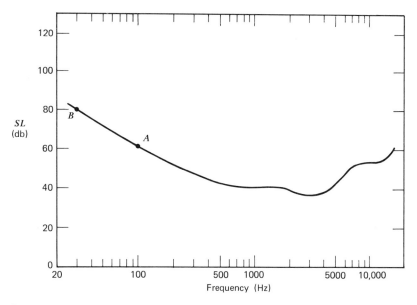

Figure 1 Equals loudness curve. $LL = 40$ phons.

For a tone of frequency f:

$$LL \text{ in phons} = SL \text{ of } 1000 \text{ Hz tone judged to}$$
$$\text{be of the same loudness}$$

We thus have a measure of loudness for all frequencies in units of phons with reference to 1000 Hz where SL in db = LL in phons.

By using the steps outlined above, we can determine the curves in Figure 2. These curves are equal-loudness curves for various loudness levels over the frequencies of musical interest. Let us look at Figure 2. The bottom curve with $LL = 0$ phons is the threshold for hearing at various frequencies. Notice that for a 100 Hz tone to be audible it must have a sound-intensity level of 37 db while a 1000 Hz tone is audible with $SL = 0$ db. In other words, if it is to be audible, a 100 Hz tone must have an intensity almost 10,000 times greater than a 1000 Hz tone. Notice from Figure 2 how the sensitivity of our ears drops off as the frequency is decreased: a 1000 Hz tone with $SL = 50$ db is piano while a 60 Hz tone with $SL = 50$ db is barely audible! Much more energy is required to make low-frequency tones subjectively as loud as those of higher frequency. This is why bass notes tend to "fade away" as we move away from the sound source and, furthermore, why high-fidelity speakers with balanced basses are very costly.

We now have a measure that enables us to know what the sound-intensity levels of two tones f_1 and f_2 must be if they are to be judged equally loud. But we do not have a basis for comparing different

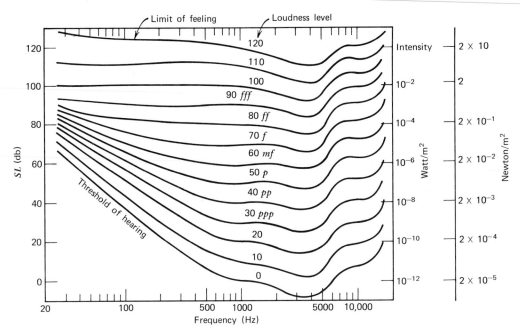

Figure 2 Equal-loudness curves. These curves are sometimes referred to as the Fletcher–Munson curves. Some loudspeaker systems have been designed with these curves in mind, thereby compensating for the low-frequency falloff in efficiency of our auditory system. (From Harvey Fletcher and W. A. Munson, "Equal-Loudness Curves," *Journal of the Acoustical Society of America*, 1933. Reprinted by permission from the Journal of the Acoustical Society of America.)

loudnesses. Some other form of measurement is needed. For example, suppose a composer is orchestrating a score for one violin. A few measures later, the composer wants to add violins so that the resulting sound will be judged "twice as loud" as the first sound. How many violins should he add?

The problem can be restated in this way: a tone with $LL = 80$ phons does *not* sound twice as loud as a tone with $LL = 40$ phons. A subjective scale of loudness is needed. The quantity *loudness* symbolized L expressed in units of *sones* is the basis for such a scale. The loudness, L, has been set up so that $L = 10$ sones is judged twice as loud as $L = 5$ sones. The relationship between loudness level (LL) and loudness (L) is shown in Figure 3. To establish the relations between LL and L, 40 phons is arbitrarily set equal to 1 sone. **Note.** An increase of the loudness level of 10 phons doubles the loudness.

What is the answer to the question concerning the number of violins to be added to one violin in order to double the subjective loudness? To double the loudness, the loudness level must be increased by 10 phons. Thus, the intensity must be 10 times greater or the sound-intensity level

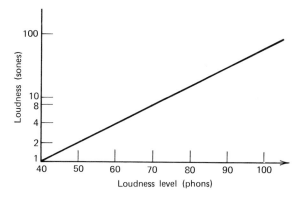

Figure 3 The experimental relation between loudness level and loudness.

must be increased by 10 db. This leads to the startling result that 10 violins will be judged twice as loud as 1 violin!

THE EFFECT OF NONLINEARITY: AURAL HARMONICS

Our auditory system has the interesting characteristic that enables us to hear more than is actually present in the exciting stimulus. Our ears literally create new tones and, furthermore, those tones that are created actually harmonize with the stimulating tone!

Let us imagine a device into which an input signal can be introduced and out of which an output signal can be extracted. Such a device is symbolized below by a box—frequently called a black box—for which the inside character is left unspecified. The input and output signals can vary in their nature by being electrical, acoustical, or whatever. The amplifier, for example, in a high-fidelity sound system is such a device; it takes a weak signal as the input and amplifies it to a strong signal.

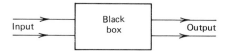

What is the difference between a good amplifier (high cost) and a mediocre amplifier (low cost)? An ideal amplifier increases the amplitude of the input signal and leaves everything else—frequency, phase, waveform—unaltered. An ideal amplifier has no distortion. A very good amplifier has very little distortion while a cheaper version may noticeably alter the input signal. Suppose, for example, the input to an amplifier is a pure tone, as illustrated below. The output signal in illustration (*a*) has the same frequency, phase, and waveform. Such an amplifier is linear. On the other hand, if the output signal resembles the one in illustration (*b*)—a clearly different waveform from that of the input—the amplifier is said to be nonlinear. A nonlinear amplifier distorts the input pure tone by

actually contributing additional pure tones. Specifically, if the frequency of the input pure tone is f, tones of frequency $2f$, $3f$, $4f$, etc. will be added by a nonlinear amplifier.

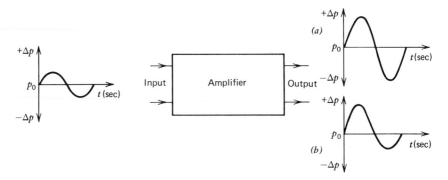

For soft sounds the human auditory system behaves in an essentially linear fashion. For loud tones, however, our auditory system becomes nonlinear. If a pure tone of frequency f is played softly, only the tone of frequency f is perceived. As the intensity of the pure tone f is increased, however, additional tones are heard corresponding to frequencies $2f$, $3f$, $4f$, etc. These additional tones are called *aural harmonics* and are a direct consequence of the ear's nonlinearity.

References

1. Roederer, Juan G., *Introduction to the Physics and Psychophysics of Music*, Springer-Verlag, 1973.

This book deals with perceptual aspects of musical sound in some detail. Professor Roederer is actively involved in research on the neuropsychological mechanisms responsible for the processing of acoustical information. This book would nicely complement the present text.

2. Wightman, Frederic L. and Green, David M., "The Perception of Pitch," *American Scientist* **62**, 208 (1974).

For Further Thought

1. Logarithms again. Suppose the intensity of a sound is 4×10^{-6} watt/m^2, what is the sound-intensity level?

$$SL = 10 \log \frac{I}{I_0} = 10 \log \frac{4 \times 10^{-6}}{1 \times 10^{-12}}$$

$$SL = 10 \log 4 \times 10^6 = 10(\log 4 + \log 10^6)$$

$$= 10(0.602 + 6)$$

$$= 10(6.602)$$

$$= 66 \text{ db}$$

Notice that the log of the product 4×10^6 is the sum of log 4 (Table 2 in Chapter 3) and log 10^6.

What is the *SL* of a sound having an intensity of 6×10^{-5} watt/m²?

2. What is the loudness level of a 1000 Hz tone whose *SL* = 60 db?

3. What is the loudness level of a 7000 Hz tone whose *SL* = 60 db?

4. What is the loudness level of a 100 Hz tone whose *SL* = 100 db?

5. One flute is playing a tone of frequency 300 Hz with a *SL* of 40 db.
 (a) What is the *LL*?
 (b) If three more flutes join the first one and play the same note, each flute contributing a sound-intensity level of 40 db, what will be the *LL* of the combination?

6. What is the loudness of a tone whose loudness level is 60 phons?

7. The loudness level of a 1000 Hz tone is 80 phons at a distance of 3 meters from the source. Assuming the validity of the inverse-square law, what would be the *LL* at a distance of 6 meters from the source?

Two Pure Tones Together

A musical sound, whether it originates with a trumpet or an oboe, consists of the superposition of *many* pure tones. Imagine a choral group: the vocal chords of each individual are set into oscillation, thereby initiating a disturbance that propagates through the air. The spreading disturbance from one voice passes through the same air space as the disturbance from a second voice, a third voice, and a fourth voice until the air is filled with the simultaneous disturbances from all the voices. Each disturbance is superimposed upon another, yet each disturbance traverses the same air space independent of all others. The resultant disturbance in any small air space is the superposition of all individual disturbances present in that air space.

The pressure variations in the air faithfully communicate the oscillations of the vocal chords to our ears. Our auditory system then takes the resultant disturbance and analyzes it back into its individual pure-tone components.

5 The Superposition of Vibrations

A loudspeaker has a cone free to vibrate. If a pure tone is heard coming from a loudspeaker, it is because the cone is vibrating with simple harmonic motion. This motion initiates a sound wave by the successive compression and rarefraction of the contiguous air. The pressure variation at any point in the path of this sound wave is described by a sine curve.

How does the speaker cone move when two pure tones are heard emanating from the loudspeaker? This is the question we shall answer and, when we have answered it for *two* pure tones, we shall have answered the same question for *any* arbitrary number of pure tones. But first it will be necessary to understand the concept of phase—a useful idea when comparing two periodic motions.

SIMPLE HARMONIC MOTION— A BRIEF REVIEW

Recall that an object oscillating with SHM is under the influence of a restoring force that increases in proportion to the amplitude of the motion. The restoring force constitutes a dynamic command that tells the oscillator exactly what to do and when to do it. Let us once again consider a mass suspended on a spring. If the mass is displaced vertically from its equilibrium position by a few centimeters and then released, it will oscillate up and down with SHM.

**COMPARING
TWO PERIODIC
MOTIONS:
PHASE**

Imagine two identical masses, each of which are supported by identical springs. If both masses are displaced +5 cm from their equilibrium positions and released, they oscillate with equal frequencies and equal amplitudes. Their motions are identical. The two periodic motions are said to be *in phase*

Once again let us displace each mass 5 cm from its equilibrium position; only this time let one mass be released from +5 cm and the other from −5 cm. The two masses again oscillate with equal frequencies and equal amplitudes; however, the motions are different: one is going up while the other is going down. In this case the two motions are completely *out of phase*. Specifically, they are ½ cycle out of phase. See the illustrations and graphs below for these two cases.

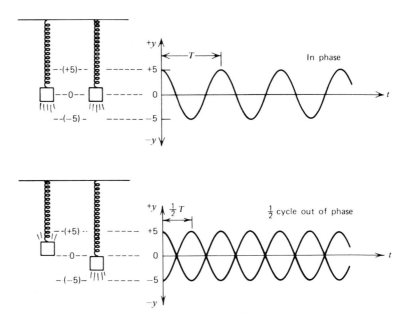

When two periodic motions are in phase or ½ cycle out of phase, the situation is clear-cut. The distinction between a *phase difference* of ¼ cycle and ¾ cycle is less obvious. For this reason it is common to speak of *relative phase*. One periodic motion is selected to serve as the reference for the other. Two examples are illustrated below.

Phase differences are usually expressed in degrees. The motion of a point traveling at constant speed along the circumference of a circle is periodic. If a line is drawn from the point on the circumference to the center of the circle, an angle, $<\phi$, is included between this line and the positive reference axis (see figure below). When the point crosses the reference axis, the angle ϕ is equal to zero as shown in (*a*). (See the following illustration.) When the point crosses the upper vertical axis (*b*), the point has completed ¼ of its cycle and $\phi = 90°$. From (*c*) and (*d*) it can

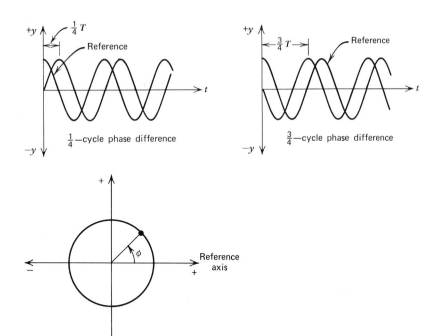

be observed that 180° and 270° correspond to $\frac{1}{2}$-complete and $\frac{3}{4}$-complete cycles, respectively. When $\phi = 360°$ (e), the point is back to the reference axis beginning a new cycle of its motion. From these considerations we can now relate phase differences with angles. This has been done in the table below. From now on we can say 180° out of phase (or a 180° phase difference) instead of $\frac{1}{2}$ cycle out of phase.

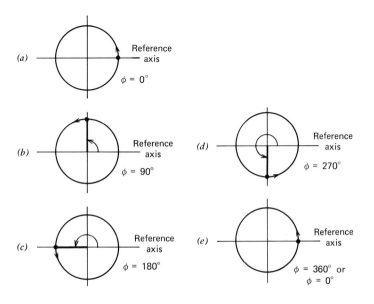

Thus far we have compared periodic motions having equal amplitudes and frequencies, but differing in their relative positions in their respective cycles. This difference we have identified as a difference in phase. We can also compare periodic motions whose amplitudes or frequencies are different. Some examples are diagrammed below.

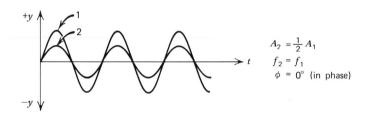

$A_2 = \frac{1}{2} A_1$
$f_2 = f_1$
$\phi = 0°$ (in phase)

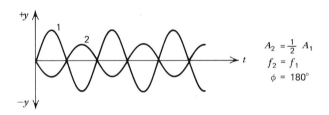

$A_2 = \frac{1}{2} A_1$
$f_2 = f_1$
$\phi = 180°$

$A_2 = \frac{1}{2} A_1$
$f_2 = 2f_1$
$\phi = 0°$

$A_2 = 2A_1$
$f_2 = 2f_1$
$\phi = 180°$

Phase relationships between two oscillations

Table 1

Phase Difference	Corresponding Angle
In phase	0°
$\frac{1}{4}$ cycle	90°
$\frac{1}{3}$ cycle	120°
$\frac{1}{2}$ cycle	180°
$\frac{3}{4}$ cycle	270°

SIMULTANEOUS VIBRATIONS— SUPERPOSITION

Now we can answer the question posed earlier: how does a speaker cone move when two pure tones are heard coming from the loudspeaker? The answer depends on the amplitudes and frequencies of the two SHMs responsible for the pure tones and the phase relationship between them. Let us assume that the two periodic motions have equal amplitudes, equal frequencies, and are in phase. Each pure-tone signal coming into the speaker constitutes a dynamic command that tells the speaker cone how to vibrate: how far to move in or out. Since the two SHMs are in phase, the dynamic commands are also in phase; that is, when command 1 says move out 0.1 mm, command 2 is simultaneously saying move out 0.1 mm. The speaker cone obeys both commands and moves out the sum 0.1 mm + 0.1 mm = 0.2 mm. When command 1 says move in 0.1 mm, command 2 is also saying move in 0.1 mm. *Result*: the cone moves in 0.2 mm. The curves below show the amplitude variation of each SHM (remember, since they are in phase, their sine curves are identical) and the resultant or *superposition curve* of the two. The superposition curve describes the actual motion of the speaker cone.

Now suppose the two identical SHMs are 180° out of phase. How does the speaker cone vibrate? When one command is saying move *out* 0.1 mm, the other command is saying move *in* 0.1 mm; since the cone is obeying both commands, it doesn't move at all! The speaker cone follows the superposition curve in the figure below and remains stationary.

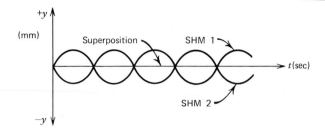

From these two examples it is apparent that one can add the amplitudes of the two SHMs algebraically to obtain the superposition curve. The addition process is shown in the following illustration with small portions of two sine curves given.

Notice from the figure:

$$6.0(\text{Point } A) = 4.0(\text{Point 1}) + 2.0(\text{Point 2})$$
$$-1.0(\text{Point } B) = +0.3(\text{Point 3}) + (-1.3)(\text{Point 4})$$
$$-6.0(\text{Point } C) = -1.0(\text{Point 5}) + (-5.0)(\text{Point 6})$$
$$+0.2(\text{Point } D) = +1.6(\text{Point 7}) + (-1.4)(\text{Point 8})$$

Each point on the superposition curve is determined in similar fashion.

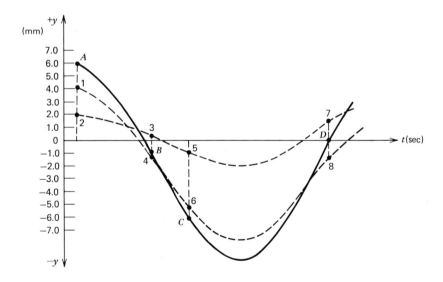

The superposition curve discussed so far describes the amplitude variation of the net vibration that results when two forces act simultaneously on the same oscillator. We need not limit ourselves to only two forces. The superposition principle is very general and can be extended to any number of forces driving an oscillator. The amplitude of the

superposition curve is simply the algebraic sum of the amplitudes of the individual oscillations.

SOME SUPERPOSITION CURVES OF MUSICAL INTEREST

Let us consider two periodic forces of frequencies f_1 and f_2 driving a speaker. We shall consider three examples.

Example 1 $f_1 = f_2$ (Unison) The amplitude of the resulting oscillation ranges from the *sum* of the two amplitudes when f_1 and f_2 are in phase to the *difference* of the two amplitudes when there is a 180° phase difference. The frequency of the resulting oscillation is the same as the component frequencies regardless of the phase relationship between them. (See question 2 at the end of the chapter.)

Example 2 $f_2 = 2f_1$ (Octave) Examine the following graphs. In each case the superposition curve is drawn with the dotted line. It is apparent that the shape of the superposition curve is dependent on the phase between f_1 and f_2. A speaker cone following any one of these four curves produces a sound perceived as an octave.

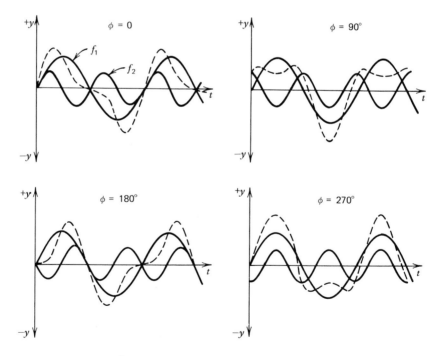

Example 3 $f_2 = \frac{3}{2}f_1$ (Fifth) After the unison and the octave, the fifth is the most widely employed musical interval. The superposition curve is shown on the next page for $\phi = 0$. You might determine the other superposition curves of the musical fifth for different relative phases.

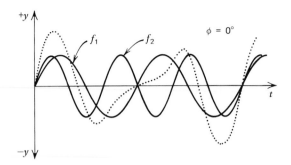

WHAT IS PURE
ABOUT A PURE
TONE?

In Chapter 2 it was indicated that the adjective "pure" had significance. Do you know what the significance is? The waveform illustrated in (a) below is not a pure tone because it can be decomposed (or analyzed) into the two "pure" tones illustrated in (b).

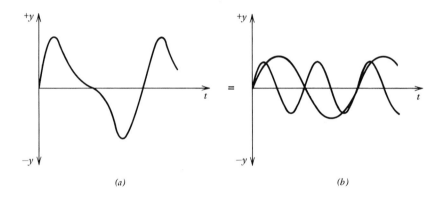

(a) (b)

Neither of the waveforms in (b) can be further broken down into simpler acoustical entities. For this reason the adjective pure is appropriate.

For Further Thought

1. Two masses are suspended from identical springs. They both oscillate with equal amplitudes and frequencies. However, they are out of phase. Both masses are observed and the times are recorded when each reaches its lowermost position. These times are:

> mass 1: . . . 9.6, 10.8, 12.0, 13.2, . . . (seconds)
> mass 2: . . . 10.5, 11.7, 12.9, 14.1. . . . (seconds)

What can you say about the relative phase between these two oscillators?

2. When two identical frequencies f_1 and f_2 are superimposed, how does the pitch vary with the phase difference between the two oscillations?

3. Two tones with frequencies f_1 and $f_2 = \frac{3}{2} f_1$ together form a musical fifth. If two periodic forces with frequencies of f_1 and $f_2 = \frac{3}{2} f_1$ and amplitudes of 0.2 mm are driving an oscillator:
 (a) Show the two waveforms on a y (mm) vs. t (sec) graph when there is a 180° phase difference.
 (b) Carefully draw the superposition curve.

4. Let us assume that we have a *very* nonlinear amplifier that exhibits a bizarre behavior. When the amplifier has an input of frequency f, the output is a mixture of f and $2f$. Furthermore, in the output the amplitude of f is four times greater than the amplitude of $2f$. If the amplifier is driving a loudspeaker, sketch a curve showing how the amplitude of the speaker cone varies with time.

6 The Superposition of Waves

The superposition principle holds for two waves traversing the same medium just as it did for an oscillator under the influence of two periodic forces. For example, suppose that two pulses are traveling along a rope—one pulse to the right and the other to the left. What happens when the two pulses meet? At the instant the centers of the two pulses coincide, a resultant pulse is formed whose amplitude is the algebraic sum of the two component amplitudes. After the moment of coincidence the two pulses continue on their way unaltered.

Sound waves behave similarly. The successive compressions and rarefactions that emanate from two sound sources spread through the

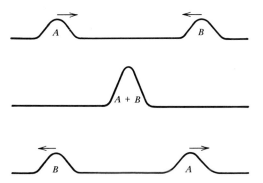

same air space. The amplitude of the resultant sound wave at a particular spatial point is the algebraic sum of the two component amplitudes. The procedure for finding the superposition curve for superimposed waves is the same as the one you learned for superimposed vibrations. Note that there is a formal difference between waves and oscillations: we have seen that the graphs representing the oscillations of a sound source show the *displacement from equilibrium* plotted against *time* while the graph for a propagating wave can take two forms. First, *pressure amplitude* can be plotted against *space*. Such a graph can be likened to a snapshot giving pressures at different spatial points at a particular instant of time. We may also plot *pressure amplitude* against *time*. This gives us a moving picture of the pressure variations at *one* spatial point.

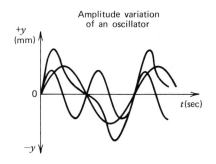

Amplitude variation of an oscillator

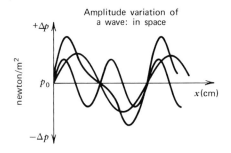

Amplitude variation of a wave: in space

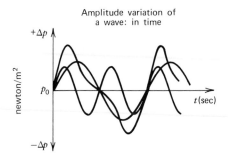

Amplitude variation of a wave: in time

WAVE SUPERPOSITIONS OF PARTICULAR INTEREST

Musical Intervals

You may have already concluded that the pressure variation of a wave is directly related to the displacement variation of the source from which the wave originates. A loudspeaker sounding a musical fifth has signals of f and $\frac{3}{2}f$ coming into it. The loudspeaker cone follows a curve like that illustrated in the previous chapter (if $\phi = 0°$) or like that below (for $\phi = 90°$). The pressure–amplitude variation of the resulting wave has the same form.

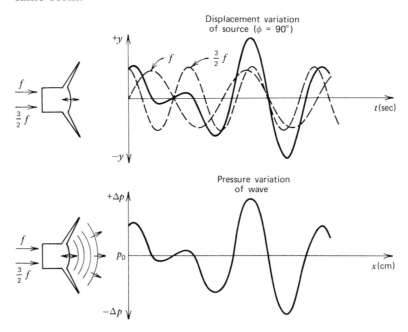

Two Sources Near Unison:
$f_1 \approx f_2$

The superposition of two pure tones whose frequencies are almost equal gives rise to an interesting effect familiar to most musicians. This time imagine the sound waves to originate from two sources as illustrated. The waves spread through the same space. If a microphone or an ear is located at point P, what is detected? To answer this question, we shall draw two sine waves whose frequencies are nearly equal. From them, we shall determine the form of the superposition curve. The figure below illustrates that at times t_A and t_C the two waves are 180° out of phase while at times t_B and t_D they are in phase. Since the amplitudes of the two sine waves are equal ($A_1 = A_2$), the amplitude of the superposition curve changes in periodic fashion from zero ($A_2 - A_1$) to a maximum of $A_1 + A_2$. It is apparent from the figure that the superposition curve is characterized by *two* frequencies: a low-frequency variation in the amplitude (dotted line) and a high frequency essentially equal to the two component frequencies f_1 or f_2 (solid line). The high frequency determines the pitch and is equal to $\dfrac{f_1 + f_2}{2}$, which is the average of f_1 and f_2.

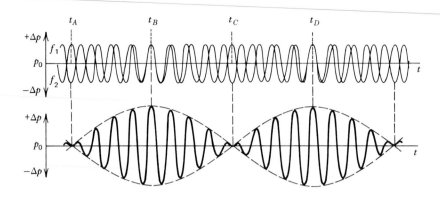

The periodic change of amplitude has a much larger period, or equivalently, a much smaller frequency. The amplitude varies with a frequency of $f_{amp} = \dfrac{f_2 - f_1}{2}$. In other words, if $f_1 = 440$ Hz and $f_2 = 444$ Hz, $\bar{f} = 442$ Hz and $f_{amp} = 2$ Hz. As we shall see in the next chapter, musicians utilize these considerations in tuning their instruments.

Standing Waves

Let us imagine a string fastened securely at each end. If the string is plucked, a disturbance propagates along the string in both directions. Reflection occurs at the string ends with the result that waves are soon moving along the string in both directions at the same time. The shape of the string at any instant is determined by the superposition of these oppositely traveling waves. Under certain conditions, the superposition of the rightward and leftward moving waves can take the interesting form referred to as a standing wave. Normally, the existence of wave implies periodic motion of a medium; however, with a standing wave some portions of the medium don't move at all! Let's see how it works.

Rather than plucking, imagine the string to be disturbed periodically with frequency f. Soon after the initiation of the disturbance, waves of frequency f and amplitude A are traveling in both directions along the string. Whether or not a standing wave results from the superposition of the oppositely traveling waves depends on the relationship between (1) the speed, v, with which the waves move along the string, (2) the length, L, of the string, and (3) the period, T, of the waves. Specifically, if a wave travels a distance of $2L$ ($2L$ is one round trip on the string) or travels a distance equal to an integral quotient of $2L$ ($2L/n$ where $n = 1, 2, 3, \ldots$) and if the wave travels one of these distances in a time equal to its own period, certain points along the string behave as though they were anchored and do not undergo vibration. The wave then appears to stand on the string.

Figure 1 shows the basic features of a standing wave. In (a) two waves

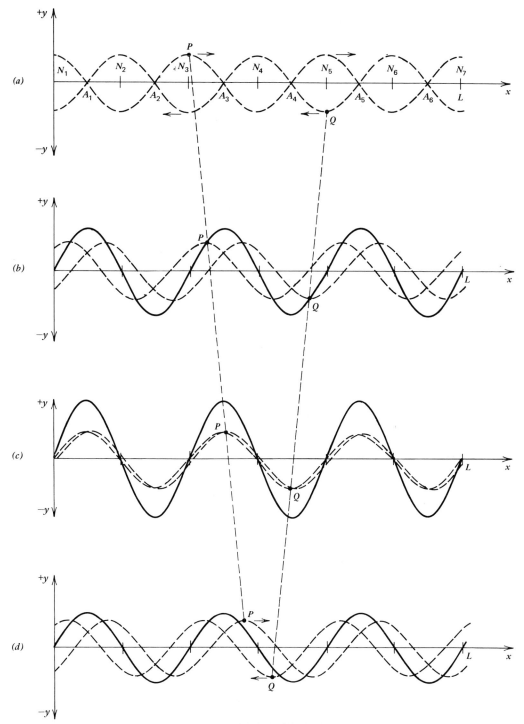

Figure 1 The creation of standing waves by two oppositely traveling waves.

are illustrated: one moving to the right (point P) and the other to the left (point Q). At this instant the string is in its equilibrium position; however, points A_1, A_3, and A_5 are moving up, points A_2, A_4, and A_6 are moving down, but points N_1 to N_7 are stationary. Figures $1b$, $1c$, and $1d$ show the string at successively later times, and it is seen that the points labeled by a N do not move while those labeled with an A move up and down. The stationary points are called *nodes*. In between each node is an *antinode*. A standing wave on a string resembles the one illustrated below.

It should not be assumed that standing waves are peculiar to strings. We shall soon see that the superposition of sound waves can also give rise to a standing wave.

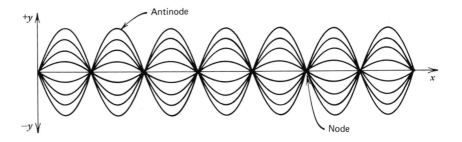

For Further Thought

1. Two pure tones with frequencies of 220 and 228 Hz are played together. A microphone is placed where waves from both sources are detected.
 (a) What is the time interval between *successive* compressions?
 (b) The amplitudes of successive compressions are different. What is the time interval between compressions having maximum amplitudes?

2. Answer question 1 for the case where the two frequencies are 212 and 220 Hz.

3. Consider a violin string of length 0.316 m (31.6 cm). Waves travel on this string with a velocity of 277 m/sec.
 (a) What is the largest period a wave can have if it is to be accommodated by the string as a standing wave?
 (b) Waves of other periods can also exist as standing waves on the string. What are some of these periods?

4. Compare the figures on pages 27 and 64. Can you express the wavelength of a standing wave in terms of the distance between nodes? In terms of the distance between antinodes?

5. The internodal distance of a standing wave on a string is 0.45 m. The frequency of the wave is 800 Hz. With what velocity does a wave propagate along the string?

6. A string is 0.80 m long. It is fastened securely at both ends: at $x = 0$ m and at $x = 0.80$ m.
 (a) Are the points $x = 0$ m and $x = 0.80$ m vibrational nodes or vibrational antinodes?
 (b) A wave with $\lambda = 0.20$ m forms a standing wave on the string. Is the point $x = 0.40$ m a vibrational node or antinode? Sketch the standing wave pattern.

7 The Perception of Two Pure Tones

We have seen

1. That the cone of a speaker with an input of two pure tones moves such that the displacement of the cone is described by the superposition curve of the two component tones.
2. That a disturbance is thereby established in the air and the resulting pressure variations are described by a superposition curve having the same form as the superposition curve describing the displacement of the source.

The air in the auditory canal is likewise disturbed, which leads to an oscillation of both the eardrum and the bone chain of the middle ear. The same superposition curve describes the amplitude variation of both the eardrum and the bone structure of the middle ear, provided that we are dealing with reasonable sound intensities so that the ear behaves linearly. Thus the air faithfully communicates the vibration of the source to the ear.

The situation changes when the disturbance enters the inner ear. There a complex tone is analyzed into its pure-tone components. In the liquid medium of the inner ear, each pure-tone component is thought to be propagated independently. What do we perceive? The answer depends both on the frequencies of the two pure tones and on the individual perceiver. The following considerations are the results of psychological testing and represent the response of the "average" auditory system.

PERCEPTUAL FEATURES OF TWO PURE TONES

Consider two pure tones whose frequencies are identical; that is, $f_1 = f_2$, or $f_2 - f_1 = 0$. In this case the two tones are said to be in unison.

We shall now consider the perceptual changes that occur when the two frequencies are *not* in unison. Let us imagine f_1 to be fixed and f_2 to slowly increase. As f_2 increases, the difference $f_2 - f_1$ also increases. This difference will be represented by Δf (read delta f) where the Greek letter Δ (delta) symbolizes difference.

Region I: Δf between 0 and 15 Hz

In the previous chapter we examined the superposition curve for two pure tones whose frequencies are nearly the same. The amplitude of this superposition curve undergoes a slow variation with a frequency designated f_{amp} where $f_{amp} = \frac{1}{2}(f_2 - f_1)$. With the variation of amplitude there is a corresponding variation in loudness. Each time the amplitude goes through a maximum, the loudness goes through a maximum. The variation in loudness is referred to as *beats*. During each cycle of the amplitude variation there are two beats arising from the two amplitude maxima. Since $f_{amp} = \frac{1}{2}(f_2 - f_1)$, the *beat frequency*, Δf_B, can be written

$$\Delta f_B = f_2 - f_1$$

As Δf_B increases, the number of beats each second increases. Near $\Delta f_B = 15$ Hz, the beats are so close together—15 beats each second—they merge together and we are no longer able to "hear out" the individual beats. When two instruments are in tune, $f_1 = f_2$ and the beats are no longer heard since $\Delta f_B = f_2 - f_1 = 0$. In tuning instruments, the object is to tune out the beats.

You might ask: what is the pitch of the beat stimulus? When two pure tones are very close in pitch, we do not hear two individual tones; rather, we perceive a *fused tone*. The pitch of the fused tone is given by

$$\bar{f} = \frac{f_1 + f_2}{2}$$

which is the pitch given by the average of the two individual frequencies.

In summary, Region I, $0 < \Delta f \leq 15$ Hz, is characterized by beats perceived at a fused-tone pitch. We shall refer to Region I as the beat region.

Region II: Δf between 15 Hz and Δf_D

As Δf increases beyond 15 Hz, a pitch having the fused-tone frequency is perceived. This tone, however, is still influenced by the beats. While individual beats cannot be perceived, the beats bestow upon the perceived sound a distinctly harsh character.

Region II, $15 < \Delta f < \Delta f_D$, is characterized by roughness in tone quality at the fused-tone pitch.

Boundary I:
$\Delta f = \Delta f_D$

Our auditory system is unable to distinguish between two tones when their frequencies are very close together. As Δf increases, however, a frequency separation is reached where we suddenly hear two separate tones. This frequency difference defines the *limit of frequency discrimination*, Δf_D. Δf_D gives the necessary difference between two frequencies for them to be heard separately. The diagram below shows the regions discussed thus far as well as the boundary giving Δf_D.

Notice from the diagram that Δf_D changes with frequency. When f_1—the fixed frequency—equals 500 Hz, Δf_D is approximately equal to 30 Hz. Thus, f_2 must equal 530 Hz before f_1 and f_2 are heard as two separate tones. If f_2 equals 524 Hz, f_1 and f_2 will be perceived as a harsh fused tone having a frequency of 512 Hz. On the other hand, if $f_1 = 2000$ Hz, then $\Delta f_D = 200$ Hz; therefore, f_2 must equal 2200 Hz if f_1 and f_2 are to be heard individually.

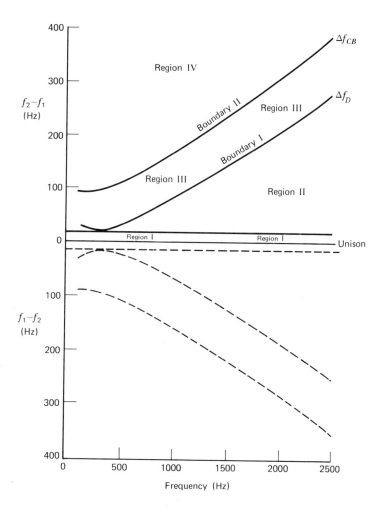

Below Boundary I two tones are perceived as a fused tone. At or above Boundary I, both tones are heard separately. Roughness still persists.

Region III:
Δf between
Δf_D and Δf_CB

In Region III, $\Delta f_D \le \Delta f < \Delta f_{CB}$, two separate tones are perceived; however, the roughness typical of Region II still persists. As before, the roughness is due to beats.

Boundary II:
Δf = Δf_CB

At a particular value of Δf the roughness fades out and two smooth tones are heard. The value of Δf at which roughness disappears defines Δf_{CB}, the *critical band*. Once again, Δf_{CB}, like Δf_D, is frequency dependent. At a frequency of 500 Hz, $\Delta f_{CB} = 110$ Hz, at 1500 Hz, $\Delta f_{CB} = 230$ Hz. The higher the frequency, the broader the critical band.

Region IV:
Δf Greater
Than Δf_CB

When $\Delta f > \Delta f_{CB}$, two smooth tones are heard. The critical band defines the separation that must exist between two frequencies in order for them to be heard individually and to have a pleasing, smooth quality. Inside the critical band a variety of perceptual difficulties are encountered.

We could now repeat our examination of the perceptual aspects of two pure tones only this time slowly *decrease* f_2 while f_1 remains fixed. We would then pass through similar regions and boundaries as we did when $f_2 > f_1$. Since the diagram is symmetric with respect to the unison line, the vertical axis can be labeled either $f_2 - f_1$ (if $f_2 > f_1$) or $f_1 - f_2$ (if $f_1 > f_2$).

COMBINATION
TONES:
NONLINEARITY
REVISITED

Aural harmonics result from the nonlinearity of our auditory system. Frequencies of $2f$, $3f$, etc. will be heard to accompany a pure tone of frequency f when it is played such that the sound intensity level at the ear is above 75 db. When *two* pure tones are played together at high intensity levels, the ear's nonlinearity creates *combination tones* that are heard in addition to the two exciting tones.

There are many combination tones that can be heard; however, some are heard more easily than others. The one most easily perceived is the difference frequency which we shall symbolize, f_{C1}:

$$f_{C1} = f_2 - f_1$$

(In our discussion we shall assume that $f_2 > f_1$.) Suppose, for example, that our two loud pure tones form a musical fifth. For a fifth $f_2 = \frac{3}{2}f_1$; therefore, the first combination tone frequency is

$$f_{C1} = f_2 - f_1 = \tfrac{3}{2}f_1 - f_1 = \tfrac{1}{2}f_1$$

Thus, when a musical fifth is played loudly, a tone will be perceived one octave below the lower frequency member of the combination.

The next two combination tones heard most easily are f_{C2} and f_{C3}

where

$$f_{C2} = 2f_1 - f_2$$

and

$$f_{C3} = 3f_1 - 2f_2$$

Again, taking the musical fifth as our example, we can find the frequencies of f_{C2} and f_{C3}.

$$f_{C2} = 2f_1 - f_2 = 2f_1 - \tfrac{3}{2}f_1 = \tfrac{1}{2}f_1$$
$$f_{C3} = 3f_1 - 2f_2 = 3f_1 - 2(\tfrac{3}{2}f_1) = 0$$

The combination tone f_{C2} is, like f_{C1}, an octave below f_1. f_{C3} would not be perceived in this case. The combination tones of other musical intervals are given in Table 1. Notice that for the octave f_{C2} and f_{C3} are not perceived while f_{C3} is not heard for either the fifth or the major sixth.

Table 1 Combination Tones for Some Musical Intervals

Interval	f_1	f_2	f_{C1}	f_{C2}	f_{C3}
Octave	f_1	$2f_1$	f_1	—	—
Fifth	f_1	$\frac{3}{2}f_1$	$\frac{1}{2}f_1$	$\frac{1}{2}f_1$	—
Fourth	f_1	$\frac{4}{3}f_1$	$\frac{1}{3}f_1$	$\frac{2}{3}f_1$	$\frac{1}{3}f_1$
Major third	f_1	$\frac{5}{4}f_1$	$\frac{1}{4}f_1$	$\frac{3}{4}f_1$	$\frac{1}{2}f_1$
Major sixth	f_1	$\frac{5}{3}f_1$	$\frac{2}{3}f_1$	$\frac{1}{3}f_1$	—
Minor third	f_1	$\frac{6}{5}f_1$	$\frac{1}{5}f_1$	$\frac{4}{5}f_1$	$\frac{3}{5}f_1$

There are a number of other combination tones. In fact, linear combinations of f_1 and f_2 of the form $f_1 + f_2$, $2f_1 + f_2$, $f_1 + 2f_2$, $f_1 + 3f_2$, $3f_1 + f_2$, $f_1 - 2f_2$, etc., are all combination tones. Most of these are very difficult to hear unless the two primary tones are very loud. Many of these combination tones, however, become important in the context of sound reproduction. As we shall discover later, poor quality sound system components such as speakers, amplifiers, and tuners behave in a non-linear fashion with the result that unwanted harmonics and combination tones are created.

In summary, when two tones f_1 and f_2 are played together *loudly*, the following tones are perceived:

Coming to our ears → f_1 , f_2

$2f_1$ $3f_1$ Aural harmonics
$2f_2$ $3f_2$ created by our auditory systems because of nonlinearity

$f_2 - f_1$ Combination tones
$2f_1 - f_2$ created by our
$3f_1 - 2f_2$ auditory system because of nonlinearity

Others

To convince yourself of the reality of these created tones, try this experiment. Play a tone f_1 loudly so that aural harmonics are created. Now play a second tone softly and vary the frequency. When the second frequency gets within a few hertz of $2f_1$, you will hear beats. Why? The aural harmonic $2f_1$ created by our auditory system together with the second soft tone form beats.

FUNDAMENTAL TRACKING

Let us again consider a musical fifth. The diagram below shows the frequencies f_1 and $f_2 = \frac{3}{2}f_1$ as well as the superposition curve, f_0. Look carefully at the three curves: in the same time that f_1 goes through *two* cycles, f_2 goes through *three* complete cycles, and the superposition curve, f_0, goes through only *one* cycle. In other words, $f_0 = \frac{1}{2}f_1$.

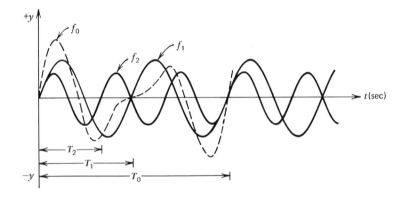

Our auditory system is sensitive to the frequency, or repetition rate, of the superposition curve. We have seen that when f_1 and $f_2 = \frac{3}{2}f_1$ are played together *loudly*, a tone of frequency $\frac{1}{2}f_1$ is perceived. This is a combination tone and arises because of nonlinear behavior of the auditory system. However, even if f_1 and $f_2 = \frac{3}{2}f_1$ are played together *softly*, a tone of frequency $\frac{1}{2}f_1$ is perceived, but in this case the auditory system's nonlinearity is clearly not involved. Rather, the repetition rate of the superposition curve is being perceived and is apparently a consequence of the way information is processed at the neurological level.

A tone of frequency f has harmonics $2f$, $3f$, $4f$, $5f$, etc. Within this family of frequencies, $2f$ and $3f$ together are a musical fifth and f, the fundamental, is one octave below $2f$. When tones f_1 and $f_2 = \frac{3}{2}f_1$ are played together, the frequency $\frac{1}{2}f_1$ can be considered as the "missing fundamental." Our auditory system places significance on this "missing fundamental" and the process of perceiving it is called *fundamental tracking*.

Following this same line of reasoning, what would you predict to be perceived when a musical fourth is played softly? In a fourth we have f_1 and $f_2 = \frac{4}{3}f_1$. If you make a drawing, you will find that the superposition curve has a repetition rate of $\frac{1}{3}f_1$ and we perceive a tone corresponding to this frequency. The frequency $\frac{1}{3}f_1$ is a twelfth (octave plus a fifth) below f_1. In a harmonic family of frequencies, $3f$ and $4f$ together form a fourth and the fundamental f is $\frac{1}{3}$ that of $3f$. Thus, the superposition curve of a musical fourth has a repetition rate corresponding to the "missing fundamental." Table 2 gives the frequencies of the "missing fundamental" for other musical intervals. You ought to verify the results expressed there.

The "missing fundamentals" of these musical intervals are not all heard with equal facility. It is easier to hear the missing fundamental for a fifth than it is for a fourth. In fact, as the frequencies of the two primary tones approach each other, perception of the missing fundamental becomes more difficult. If you sketch the superposition curves for various intervals, you will find that its complexity increases as you go from a fifth, to a fourth, to a third, etc. Apparently, the human auditory system has more difficulty tracking when the superposition curve is complicated.

Two-tone stimulus

Missing fundamental

Two-tone stimulus

Missing fundamental

Table 2 Frequencies of the "Missing Fundamental"

Musical Interval	f_1	f_2	Frequency of "Missing Fundamental"
Octave	f	$2f$	f
Fifth	f	$\frac{3}{2}f$	$\frac{1}{2}f$
Fourth	f	$\frac{4}{3}f$	$\frac{1}{3}f$
Third	f	$\frac{5}{4}f$	$\frac{1}{4}f$
Sixth	f	$\frac{5}{3}f$	$\frac{1}{3}f$

For Further Thought

1. Two tones having frequencies of 512 and 516 Hz are played to-gether.
 (a) What is f_{amp}?
 (b) What is Δf_B?
 (c) Are beats perceived?

2. An oboe is playing a 440 A. A second oboe plays the same note and six beats are heard each second. If we assume the first instrument's tone has a frequency exactly equal to 440 Hz:
 (a) What is the frequency of the second oboe?
 (b) Is the second oboe sharp or flat with respect to the first oboe?

3. A pure tone with frequency 1000 Hz is played. A second pure tone, higher in pitch, is played with the first tone. If two distinct tones are perceived, what is the smallest possible frequency of the second tone?

4. Two pure tones having frequencies of 1000 and 1040 Hz are played together softly. What is perceived?

5. Two pure tones having frequencies of 1500 and 1550 Hz are played together softly. What is perceived?

6. Two pure tones having frequencies of 500 and 650 Hz are played together softly. What is perceived?

7. At a frequency of 750 Hz:
 (a) What is Δf_D?
 (b) What is Δf_{CB}?

8. Two pure tones having frequencies of 440 and 587 Hz are played together loudly. What are the frequencies of f_{C1}, f_{C2}, and f_{C3}, the combination tones likely to be perceived?

9. Two pure tones with frequencies 440 and 660 Hz are played together. Identify what is likely to be perceived when:
 (a) The two are played together softly.
 (b) The two are played together loudly.

10. Two tones have periods of $T_1 = 0.00455$ sec and $T_2 = 0.00303$ sec. If these two tones are played together, what is the period of the missing fundamental?

11. Two pure tones have frequencies of 49 and 61 Hz. What is the frequency of the missing fundamental?

Archetypal Instruments: The Tromba Marina, the Fife, the Coach Horn, and the Drum

In the twentieth century incredible change has occurred in practically every area of human experience. Does anyone listen to a radio, drive a car, cook on a stove, sleep on a mattress, or mow a lawn with 30-year-old implements? Changes are made so rapidly that today's innovation is tomorrow's obsolescence.

Musical instruments, however, are an exception. The clarinet, trombone, and guitar built today are basically the same as those built earlier. Older violins, moreover, such as the Stradivarius and Guarnerius are the standards by which modern violins are judged. The modern piano took its present form in 1855 when the American piano manufacturer Henry Steinway introduced the cast-iron frame in his grand piano.

LXII *Tromba Marina*

(From *The Showcase of Musical Instruments* by Filippo Bonanni, Dover, New York, 1964.)

Fife: 1100s

Coach horn: 1700s

Why hasn't there been an influx of new instruments based on new principles, featuring novel designs? The reason, I believe, is because of the peculiar and unique properties of the stretched string and the air column. Both the taut string and the air column can vibrate with a basic frequency and with whole-number multiples of that frequency. From early times, a string stretched between two pegs mounted on some kind of base or an air column defined by a hollow bamboo stalk, perhaps with finger holes along its length, have been the basis of musical sound.

The tromba marina is a simple single-stringed instrument and is a predecessor of the contemporary stringed instrument. The fife is a wind instrument in which a cylindrical column of air is the basis for sound production. Today's flute and clarinet are decendents of the fife. The coach horn, another wind instrument, has a conical air column and can be compared with the modern oboe and saxaphone.

(From *The Showcase of Musical Instruments* by Filippo Bonanni. Dover, New York, 1964.)

The drum, originating early in the activities of Homo sapiens, is basically different. A drum does not produce a family of frequencies harmonically related. Typically, percussion instruments produce frequencies that we shall call *overtones*.

In Part IV we shall focus on the very heart of musical instruments: the stretched string, an air column, and taut membranes. Our objective is to understand their behavior as the origin of musical sound. Furthermore, we shall examine how the immediate environment modifies the sound originating with the string or the air column. Finally, we shall raise some questions about the perception of complex sounds. These considerations will establish the basis for an examination of the instruments of musical sound in Part V.

8 The Origin of Complex Tones: Strings, Tubes, and Membranes

Musical instruments produce complex tones. The waveform of the sound emanating from any musical instrument appears considerably more complicated than the waveforms of the musical intervals we saw in Chapters 5 and 6. Within this complexity, however, there is order. The waveforms of musical sound are repetitive and have a definite period and frequency. A fundamental theorem in mathematics states that any periodic waveform, no matter how complicated it may be, can be created by superimposing the proper collection of sine curves. In other words, the complex sounds of music are the superposition of many pure tones.

A complex tone is characterized by the pure tones that make it up: if the frequencies, amplitudes, and relative phases of the component tones are known, the complex tone is completely specified. There are standard laboratory instruments called wave analyzers that can determine the frequencies and amplitudes of the pure tones present in a complex tone. The results of such an analysis could be arranged as in Table 1. The constituent pure tone frequencies are often related to one another; that is, the frequencies f_2, f_3, f_4, etc. are whole-number multiples of f_1. In this case, we shall refer to the pure tones as *harmonics*. There are some

complex sounds, for example, those originating with the percussion instruments, where f_2, f_3, etc. are *not* whole-number multiples of f_1 and we shall refer to such pure tones as *overtones*.

Table 1 Analysis of a Complex Tone

	Frequency (Hz)	Intensity (watt/m²)
First component	f_1	I_1
Second component	f_2	I_2
Third component	f_3	I_3
—	—	—
—	—	—
etc.	etc.	etc.

The results of analyzing a complex tone can also be represented graphically. In such a graph, the frequency is plotted along the horizontal axis and the *relative intensity* along the vertical axis. The usual practice is to assign the arbitrary value of 1.0 to the intensity of the lowest frequency component (the fundamental) and to relate the intensities of all higher frequencies to it; hence, the term relative intensity. The graph of a pure tone is particularly simple since there is only one component.

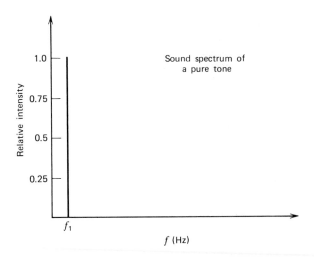

The graph of the complex tone shown below indicates that (i) the component frequencies are multiples of f_1 and are thus harmonically related; (ii) that the intensity of f_5 is $\frac{1}{2}$ that of f_1; and (iii) that f_8 is missing altogether. We shall refer to such a graph as a *sound spectrum*.

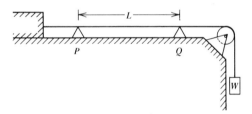

THE VIBRATIONAL MODES OF A STRETCHED STRING

The strings in musical instruments are stretched taut across two supports: a bridge and some kind of end support. The apparatus below, in which the string is supported at points P and Q, is quite analogous. The distance between points P and Q is L, which we shall consider to be the length of the string. The tension in the string is maintained by the suspended weight W.

Now the question is: what factors determine the frequency—or frequencies—with which a taut string of length L will vibrate? We can, of course, force the string to vibrate with any frequency; however, here we are interested in the natural frequencies of the string. Let us begin by recognizing that the string is an elastic medium. If it is disturbed, two things happen: first, that portion of the string that was disturbed is set into periodic motion and, second, the disturbance propagates along the string in both directions. At the end points P and Q the waves are reflected with the result that all portions of the string are vibrating as waves race back and forth along the string. With waves moving in both directions along the string, we have the potential for standing waves.

This potential will be realized under the proper circumstances, which we shall now examine.

The relationship $v = f\lambda$ holds true for all waves. Therefore, if we can determine something about the wavelength and the speed of propagation of the string waves, we shall have learned something about the frequency with which the string vibrates. The wavelengths that can be accommodated by the string are determined by the fact that points P and Q define a length of string that *can* vibrate while the points P and Q themselves *cannot* vibrate. Points P and Q are always vibrational nodes. The first vibrational mode of the string has a single antinode between P and Q. The string has a vibrational pattern like that illustrated below. For this vibration we see only one half of a wavelength at any instant and conclude that the wavelength must be $2L$. This vibration is called the fundamental or first harmonic and its frequency and wavelength are symbolized by f_1 and λ_1, respectively. The second vibrational mode has two antinodes plus a node between the ends P and Q. This, and a third vibrational mode are illustrated along with the symbols we shall associate with them. In the same figure, you should sketch the wave pattern for the fourth harmonic.

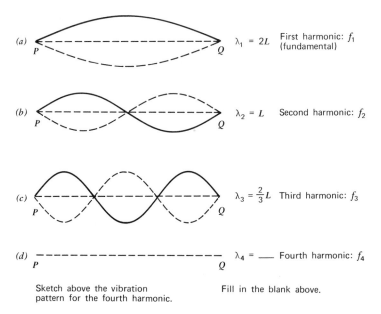

(a) P Q $\lambda_1 = 2L$ First harmonic: f_1 (fundamental)

(b) P Q $\lambda_2 = L$ Second harmonic: f_2

(c) P Q $\lambda_3 = \frac{2}{3}L$ Third harmonic: f_3

(d) P Q $\lambda_4 =$ ___ Fourth harmonic: f_4

Sketch above the vibration pattern for the fourth harmonic. Fill in the blank above.

Vibrational modes of a stretched string

From these drawings we see that there is always an integral number of half wavelengths. This fact enables us to generalize the wavelength results and make them more concise. The table below summarizes these results.

Wavelength	Wavelength in Terms of L
λ_1	$2L = \dfrac{2L}{1}$
λ_2	$L = \dfrac{2L}{2}$
λ_3	$\tfrac{2}{3}L = \dfrac{2L}{3}$

Looking at these tabulated results we are led to write

$$\lambda_n = \frac{2L}{n} \qquad n = 1, 2, 3, \ldots$$

where the wavelength of the second harmonic can be obtained by substituting $n = 2$ into the relationship and the wavelength of any other harmonic by substituting the appropriate value of n. We shall soon employ this generalized result to determine the frequencies of the vibrational modes of the stretched string.

The speed with which the waves move along the string is dependent upon properties of the string itself; namely, upon the tension of the string and upon its linear density. The greater the tension, the quicker the string snaps back toward its equilibrium position, and the faster the wave propagates along the string. On the other hand, the thicker the string, the slower the string snaps back toward its equilibrium position, and the slower the wave propagates along the string. If we symbolize tension by \mathcal{T} (expressed in force units called newtons) and the linear density by d (expressed in kilograms per meter), the speed is given by

$$v = \sqrt{\frac{\mathcal{T}}{d}}$$

The appearance of the square-root sign may be surprising, but it is necessary if we are to obtain the correct speed for waves on a string.

Now we can determine the vibrational frequencies of a stretched string. From the basic relation $v = f\lambda$ we can write

$$f\lambda = v \qquad f\lambda = \sqrt{\frac{\mathcal{T}}{d}}$$

$$f = \frac{v}{\lambda} = \frac{1}{\lambda}\sqrt{\frac{\mathcal{T}}{d}}$$

Using the symbols defined above, we can write more explicitly

$$f_1 = \frac{1}{\lambda_1} \sqrt{\frac{\mathcal{T}}{d}} = \frac{1}{2L} \sqrt{\frac{\mathcal{T}}{d}}$$

$$f_2 = \frac{1}{\lambda_2} \sqrt{\frac{\mathcal{T}}{d}} = \frac{1}{L} \sqrt{\frac{\mathcal{T}}{d}}$$

or

$$f_n = \frac{1}{\lambda_n} \sqrt{\frac{\mathcal{T}}{d}} = \frac{1}{2L/n} \sqrt{\frac{\mathcal{T}}{d}}$$

or

$$f_n = \frac{n}{2L} \sqrt{\frac{\mathcal{T}}{d}} \qquad n = 1, 2, 3, 4, \ldots$$

Since $f_1 = 1/2L \sqrt{\mathcal{T}/d}$, we can summarize the results as

$$f_n = nf_1 \qquad n = 1, 2, 3, 4, \ldots$$

in which form it is obvious that all frequencies are whole-number multiples of a fundamental frequency. Thus, a string can vibrate with a large number of frequencies, but they are harmonically related.

In Chapter 6 we discussed standing waves. The harmonic frequencies of the string obtained in the present section are the frequencies of the standing waves that the string can accommodate. The natural frequencies of the string can be written in terms of the wave's period:

$$f_n = \frac{1}{T_n} = \frac{n}{2L} v$$

This expression can be solved for T_n:

$$T_n = \frac{1}{n} \frac{2L}{v} \qquad n = 1, 2, 3, \ldots$$

In Chapter 6 it was stated that "if a wave travels a distance of $2L$ or an integral quotient of $2L$ in a time equal to its own period," a standing wave is formed. The result above symbolizes this statement. If a round trip is made in a time equal to one period of the wave ($n = 1$), the string is vibrating with its first harmonic frequency and *one* antinode is formed along the string. If, on the other hand, the wave makes one-half a round trip in a time equal to its period ($n = 2$), the string is vibrating with its second harmonic frequency and *two* antinodes are formed along the string. In these terms, what does $n = 3$ imply?

Do the results of this section make sense? To begin with, you may realize that if a guitar or violin string is tightened (tension increased), the pitch or frequency increases. This is consistent with our result. Furthermore, the low-pitched strings of an instrument are thicker and heavier (linear density greater) than the high-pitched strings. This observation is

also consistent with our results. However, you ought to try two experiments for yourself and determine whether a string can vibrate with several frequencies. For each experiment you'll need a piano.

Experiment One Slowly depress the C_2^\sharp key on the piano and hold it down. In this way, the damper is lifted from the string, but the hammer does not strike it. Now strike C_3 hard and staccato. After the initial sound has faded away, you will hear the C_2 string vibrating clearly and beautifully in its second harmonic vibrational mode. Lift the C_2 key and the damper will make contact with the string thereby terminating the sound. Repeat the above steps by striking G_3, C_4, E_4, G_4, etc.—all hard and staccato—and you will hear the C_2 string vibrating in its third, fourth, fifth, and sixth harmonic, modes, respectively. Now repeat the experiment, but this time strike D_3 or F_3. Nothing is heard on the C_2 string because neither D_3 nor F_3 are harmonics of C_2.

This experiment is convincing evidence that a string can vibrate with many different frequencies provided that they are harmonically related.

Experiment Two Again, slowly depress the C_2 key and hold it down. With your right underforearm hit all the keys between C_3 and C_5. After the cacophony of sound has subsided, you will hear the C_2 string vibrating in a full dominant seventh chord: C_3, G_3, C_4, E_4, G_4, B_4^\flat, C_5.

This experiment permits the conclusion that a string can vibrate with several of its harmonic frequencies simultaneously.

Yes, our result makes sense.

THE VIBRATIONAL MODES OF AN AIR COLUMN A column of air is like a stretched string: it can also vibrate with frequencies that are harmonically related. Consequently, when the fundamental frequency is excited, it is almost inevitable that other harmonics will also be excited, resulting in a rich, full, and pleasing sound.

The Open Cylindrical Tube Consider a cylindrical tube of length L, open at both ends. If you blow carefully across one end of the tube, a disturbance is initiated that propagates down the tube. The wave is partially reflected at the opposite end even though it is open. With proper blowing conditions, the incident and reflected waves can interfere to produce a standing wave in the tube. The tube is then said to be in resonance and a strong musical tone is the

*The lowest C on the piano is C_1. The D one whole step above C_1 is D_1. The next C, an octave above C_1, is C_2, etc. A_4 would be the A above C_4 or the A above middle C.

result. The pitch of the tone tells us the frequency with which the air inside the tube is vibrating.

The walls of the tube provide a rigidity against which the enclosed air can pile up and hence develop rather large pressure variations. At the ends of the open tube, however, there are no rigid walls against which the air can pile up; therefore, the pressure maintained is the same constant value as the external atmospheric pressure. The open ends are therefore always pressure nodes and are analogous to the fixed ends of a string. To determine the frequencies of the vibrational modes of an air column, we shall proceed as we did in our analysis of the string.

The fundamental, or first harmonic, has a pressure node at each open end and a pressure antinode in the middle of the tube. This vibrational mode and others are illustrated below.

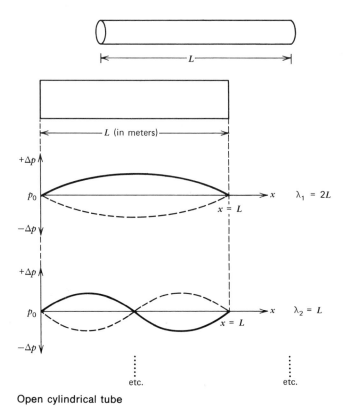

Open cylindrical tube

It should be clear from the figure that $\lambda_n = 2L/n$, $n = 1, 2, 3, 4, \ldots$ (If this result is confusing, review the corresponding analysis for the string.) The vibrational frequencies for the air column can be written as follows:

$$f_n = \frac{v}{\lambda_n} = \frac{n}{2L} v$$

but $v = 20.1\sqrt{T_A}$; therefore,

$$f_n = \frac{n}{2L} 20.1\sqrt{T_A}$$

$$f_n = n\frac{10.05}{L}\sqrt{T_A} \qquad n = 1, 2, 3, \ldots$$

where L must be expressed in meters. These are the resonance frequencies of a tube of length L. When the air in a tube is periodically disturbed at one of its resonance frequencies, a strong tone results. Disturbances initiated at other frequencies are quickly dissipated in the tube. If we assume room temperature to be 22°C or 295°K, then

$$f_n = n\frac{10.05}{L}\sqrt{295}$$

or

$$f_n = n\frac{173}{L} = nf_1$$

Like the string, the open tube can accommodate a family of harmonically related frequencies. The length of the tube determines the frequency of the fundamental and hence the frequency where the family of harmonics originates. We have assumed the length, L, to be the measured length of the tube. This assumption is almost, but not quite, valid. Theory shows, and experiment confirms, that the pressure node is slightly outside the open end, which makes the effective length longer than the actual length. The end correction is $0.346\sqrt{A}$, where A is the cross-sectional area (expressed in square meters) of the tube. Since this correction must be applied to both open ends, the effective length of an open tube is

$$L_{\text{eff}} = L_{\text{actual}} + 0.692\sqrt{A}$$

Obviously, the larger the cross-sectional area of the tube, the more significant is the correction factor. The previous results can be altered to include the end correction if the effective length is used; that is,

$$f_n = n\frac{10.05}{L_{\text{eff}}}\sqrt{T_A}$$

The Semiclosed Cylindrical Tube

A semiclosed tube, open at one end and closed at the other end, behaves differently than the open tube. The reason is that the closed end of the tube provides a rigid barrier against which pressure variations not only can develop but, in fact, do develop. Thus we have a pressure antinode at the closed end and a pressure node at the open end. In order to find the wavelengths that the semiclosed tube can accommodate, we proceed as before and fit nodes and antinodes along the length of the tube

subject to the requirement that there is a node at one end and an antinode at the other end. This has been done in the figure shown below.

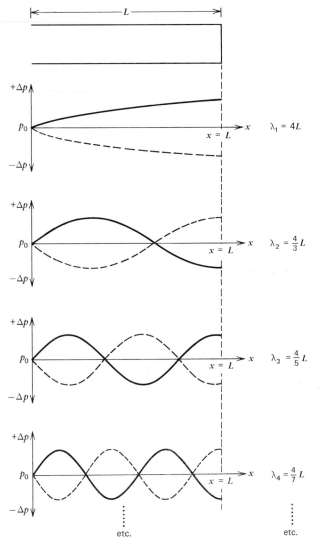

The semiclosed cylindrical tube

From the illustrations and the attending wavelength values it is apparent that the resonances of a semiclosed tube always have an integral number of quarter wavelengths. Upon closer inspection, however, we note that there are always an *odd number* of quarter

wavelengths. Some wavelengths seem to be missing. If we change our notation slightly, we can generalize our results and gain understanding of the semiclosed tube.

Original Notation	Change	Modified Notation
$\lambda_1 = 4L$	No Change	$\lambda_1 = 4L$
$\lambda_2 = \frac{4}{3}L$	Change λ_2 to λ_3	$\lambda_3 = \frac{4}{3}L$
$\lambda_3 = \frac{4}{5}L$	Change λ_3 to λ_5	$\lambda_5 = \frac{4}{5}L$
$\lambda_4 = \frac{4}{7}L$	Change λ_4 to λ_7	$\lambda_7 = \frac{4}{7}L$

In the modified form, these wavelengths can be expressed in a general formula in terms of which we can write the frequencies:

$$\lambda_n = \frac{4L}{n} \qquad n = 1, 3, 5, 7, \ldots$$

$$f_n = \frac{v}{\lambda_n} = n \frac{20.1}{4L} \sqrt{T_A}$$

$$f_n = n \frac{5.025}{L} \sqrt{T_A} \qquad n = 1, 3, 5, 7, \ldots \text{(odd)}$$

At room temperature, 23°C, this result becomes

$$f_n = n \frac{86}{L} \qquad n = 1, 3, 5, 7, \ldots \text{(odd)}$$

Once again, the effective length is slightly longer than the actual length. For one open end the effective length is given by $L_{\text{eff}} = L_{\text{actual}} + 0.346\sqrt{A}$.

We can draw two important conclusions from the results above. First, the air in a semiclosed cylindrical tube can vibrate with many harmonically related frequencies; however, the even harmonics f_2, f_4, f_6, etc. are absent. The sound emanating from a semiclosed tube has a characteristic quality and it derives that character by virtue of its odd harmonic makeup. Second, the fundamental frequency of a semiclosed pipe is an octave lower than that of an open pipe of the same length. This property is utilized in the design of pipe organs since space can be saved by employing closed pipes.

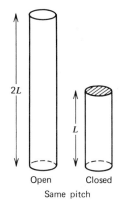

2L

Open Closed
Same pitch

Conical Tube We have seen that both the stretched string and the cylindrical air column can vibrate with a number of harmonically related frequencies; that is, they can vibrate with frequencies that are exact whole-number multiples of a fundamental frequency. It is this property that gives the taut string and the cylindrical air column their premier status as sources of musical sound.

One can imagine tubes with a variety of shapes other than cylindrical. The question to be answered is: are there any other tube shapes in which the enclosed air vibrates with frequencies that are multiples of some fundamental frequency? Two other shapes meet this condition. The first of these is a flared tube whose radius increases by a factor of 11.31 for each doubling of the distance from the small end. In other words, if two centimeters from the small end, the radius of the tube is one centimeter, then four centimeters from the small end the radius would be 11.31 centimeters and eight centimeters from the small end the radius would be (11.31 × 11.31 cm) 128 centimeters. This tube flares out so rapidly that there would be tremendous mechanical difficulties both in constructing it and handling it. Unfortunately, tubes with more gradual flares do not meet the necessary condition.

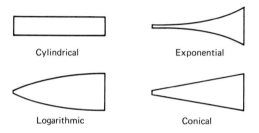

Tube shapes

There is a second shape that qualifies. It is a conical tube. This tube has straight sides meeting at a point. The air contained within a perfect cone of length L vibrates with exactly the same frequencies as the open cylindrical tube having the same length. Thus, our earlier result, $f_n = n(10.05/L)\sqrt{T_A}$, is applicable. The conical tube encountered in musical applications, however, is not complete to the tip; rather, the end is cut off. Consider a cone, whose complete length is l, with its small end removed and closed. If the length of the cut-off portion is small in comparison to the length of the complete cone—specifically, if l_0 is less than $\frac{1}{4}l$, then the vibrational frequencies are essentially the same as those of the complete cone. In other words, the conical tube, cut off and closed at the small end, also behaves like an open cylindrical tube provided that the cut-off portion is small.

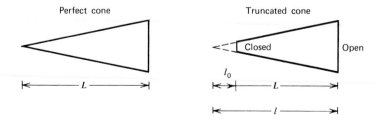

If we perform major surgery on the conical tube and cut off a large portion, the vibrational characteristics of the enclosed air changes. Consider the conical tube pictured below. I have calculated the vibrational frequencies for this tube, and they are listed, in terms of L, in Table 2. [Remember, for the open cylindrical tube $f_n = n(v/2L)$]

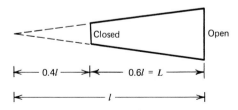

Table 2 Resonance Frequencies for Truncated Cone Where $l_0 = 0.4l$

$$f_1 = 0.75 \frac{v}{2L}$$

$$f_2 = 1.60 \frac{v}{2L} = 2.13 f_1$$

$$f_3 = 2.56 \frac{v}{2L} = 3.41 f_1$$

$$f_4 = 3.55 \frac{v}{2L} = 4.73 f_1$$

$$f_5 = 4.54 \frac{v}{2L} = 6.05 f_1$$

$$f_6 = 5.53 \frac{v}{2L} = 7.37 f_1$$

The data in Table 2 make it clear that the higher frequencies of our sample tube are not simple multiples of the fundamental frequency.

Expressing these calculated frequencies as in Table 2 obscures their character. Let us rewrite the frequencies in terms of a closed cylindrical tube for which $f_n = n(v/4L)$:

$$f_1 = 1.49 \frac{v}{4L}$$

$$f_2 = 3.20 \frac{v}{4L}$$

$$f_3 = 5.12 \frac{v}{4L} \approx 5 \frac{v}{4L} \approx f_5$$

$$f_4 = 7.09 \frac{v}{4L} \approx 7 \frac{v}{4L} \approx f_7$$

$$f_5 = 9.07 \frac{v}{4L} \approx 9 \frac{v}{4L} \approx f_9$$

$$f_6 = 11.06 \frac{v}{4L} \approx 11 \frac{v}{4L} \approx f_{11}$$

When the frequencies are expressed in this fashion, the figures suggest that this tube takes on the odd-harmonic character of a semiclosed cylindrical tube.

Let us summarize the results for the truncated conical tube. When the cut-off portion is small, the truncated conical tube behaves like an open cylindrical tube. When the cut-off portion is large, the higher frequency resonances are close in value to those of a semiclosed tube; however, the higher frequencies are overtones rather than harmonics.

THE VIBRATIONAL MODES OF A STRETCHED MEMBRANE

Suppose a membrane of uniform thickness is stretched over a circular ring of radius R in such a manner that the magnitude of the tension throughout the membrane is a constant. What are the frequencies with which such a membrane can vibrate?

The membrane is a more complex physical system than either the stretched string or the air column. This is because a disturbance initiated on the membrane can propagate in two directions. With both the string and the air column, the disturbance can only propagate in one direction. Although this additional dimension makes a thorough analysis more complex, we can proceed in the same way as we did with the string and air column.

Our analysis of both the string and the air column was based on the recognition that certain conditions must always be satisfied; for example, the ends of the string are vibration nodes, the open end of a cylindrical tube is a pressure node while the closed end is a pressure antinode. The circular membrane of radius R always has a vibrational node at the rim. This condition must always be satisfied. Furthermore, like the string, the vibrational frequencies of the membrane increases with increasing tension and decrease with increasing density; in fact, the frequencies are proportional to $\sqrt{\mathcal{T}/d}$, just as with the string. The frequency of the lowest vibrational mode is

$$f_1 = \frac{0.383}{R} \sqrt{\frac{\mathcal{T}}{d}}$$

where \mathcal{T} is the tension and d is the areal density in kilograms per square meter. The radius R is expressed in meters and \mathcal{T} in newtons.

At this point the similarity between the string and the drumhead ends. Whereas the string has a family of frequencies harmonically related, the membrane has an overtone series. Some of the vibrational modes along with their frequencies are shown on the following page. In these figures the dotted lines represent vibrational nodes. In (*a*), all parts of the membrane are vibrating up and down in phase. The amplitude is greatest at the center and zero at the rim. In (*b*), the right half goes up (+) while the left goes down (−) and vice versa. The right and left halves are 180° out of phase. The remaining figures follow the same reasoning.

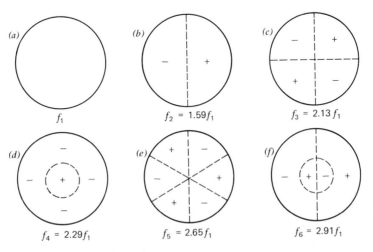

Vibrational modes of membranes

SUMMARY

When an object is forced to oscillate at one of its natural frequencies, resonance occurs. The natural frequencies—or resonant frequencies— of a stretched string, open, and semiclosed cylindrical tubes are described by similar formulae, given by

$$f_n = n\frac{1}{2L}\sqrt{\frac{\mathscr{T}}{d}} \qquad n = 1, 2, 3, \ldots$$

$$f_n = n\frac{10.05}{L}\sqrt{T_A} \qquad n = 1, 2, 3, \ldots$$

$$f_n = n\frac{5.025}{L}\sqrt{T_A} \qquad n = 1, 3, 5, 7, \ldots$$

respectively. These frequencies represent the harmonic family of the string or air columns and are whole-number multiples of the fundamental frequency. This property of strings and air columns has stimulated

the imagination of scientists, musicians, and mystics from the time of Pythagoras to the present day.

Membranes, however, are different. The resonant frequencies of membranes are overtones and are not integrally related to a fundamental frequency. Thus, there is no simple formula for the family of overtones of a membrane.

The conical tube, an important source of musical sound, is also different in that it can have either a family of harmonics, or a family of overtones. The characteristic behavior of the cone is determined by the truncated portion of the tube. When the cut-off portion is large, the higher frequencies of the tube behave like those of a semiclosed tube; however, they are overtones, not harmonics.

Strings, tubes, and membranes are the sources of musical sound. Only those frequencies that are present in their complex vibrations can be present in the transmitted sound. There are other factors, however, that influence the nature of the transmitted sound. These factors are the subject of the next chapter.

For Further Thought

1. From the following data construct a sound spectrum.

$$f_1 = \ 200 \text{ Hz} \qquad I_1 = 0.001 \text{ watt/m}^2$$
$$f_2 = \ 400 \text{ Hz} \qquad I_2 = 0.0005 \text{ watt/m}^2$$
$$f_3 = \ 600 \text{ Hz} \qquad I_3 = 0.0002 \text{ watt/m}^2$$
$$f_4 = \ 800 \text{ Hz} \qquad I_4 = 0$$
$$f_5 = 1000 \text{ Hz} \qquad I_5 = 0.0001 \text{ watt/m}^2$$

2. It might be useful to review question 8 in Chapter 3 before proceeding with this question. From the following data construct a sound spectrum.

Frequency (Hz)	Amplitude (Arbitrary Units)
150	2
300	1.5
450	0.8

3. Let us consider the violin string that formed the basis of question 3, Chapter 6. Waves travel on this string with a velocity of 277 m/sec. If the string has a linear density of 0.00065 kg/m, with what tension is it stretched?

4. A string has a linear density of 0.035 kg/m. If it is stretched to a tension of 667 newtons, what is the velocity of the wave on the string?

5. Consider the string in the previous question. If the string is 1.0 m long, what is the frequency of the fourth resonant mode?

6. A harp string is 52.0 cm long and it is tuned to a frequency of 660 Hz. Find the wavelength:
(a) Of the fifth harmonic on the string.
(b) Of the sound waves in the air (at 22°C) that results from the fifth harmonic.

7. An open cylindrical tube is 0.05 m long. If you ignore the end correction factor, what would be the frequency of the third harmonic? Assume a temperature of 10°C.

8. Two open organ pipes have lengths of 0.60 and 0.64 m. These two pipes are sounded together. If we assume the speed of sound is 345 m/sec, what is perceived?

9. A tube is sounded at room temperature. It is observed that the tube can be made to oscillate rather easily at several frequencies that include 1000, 1400, and 1800 Hz.
(a) Is the tube an open or semiclosed cylindrical tube?
(b) What is the length of the tube?

10. A semiclosed cylindrical tube has a length of 0.5 m and a radius of 6 cm. What is the frequency of this tube's first harmonic?

11. A conical tube has a length of 50 cm. The cone is cut 20 cm from the tip, thereby forming a truncated cone. The lowest frequency with which the enclosed air oscillates is 428 Hz, which occurs when the speed of sound is 345 m/sec. What other resonant frequencies does this truncated cone possess?

12. Let us assume an instrument has strings of length 76 cm. Part of the string is suspended over a fingerboard. When the string is pressed to the fingerboard:
(a) What happens?
(b) When the string is fingered a distance of 8 cm from the end, the string vibrates with a frequency of 300 Hz. Where would the finger have to be placed to obtain a vibration frequency of 311 Hz?

9 The Transmission of Complex Tones from Strings, Tubes, and Membranes

Strings, air columns, and membranes can vibrate with many frequencies simultaneously. What determines which frequencies are present in the transmitted sound? The answer is twofold. First, the nature of the transmitted sound depends very specifically on the manner in which the vibration of the string, the air column, or the membrane is initiated.

Second, it depends on further structural characteristics that we shall refer to as the immediate environment.

For example, strings are mounted on some kind of base. The vibrations of the string initiate vibrations of the base through a coupling device known as a bridge. How the base responds to these vibrations affects the nature of the transmitted sound. Tubes are flared at one end and fitted with a mouthpiece at the other end. The bell of an instrument not only affects the intensity of the transmitted sound, it also, depending on the geometry, influences the timbre of the transmitted sound. The mouthpiece exerts a determining influence on the resonant frequencies set up in the tube of the instrument. Finally, the shape and size of the structure over which a membrane is stretched influences the character of the sound wave initiated.

THE GENERATION OF COMPLEX VIBRATIONS ON A STRING

There are three methods commonly employed by musicians to set a string into vibration. These three techniques are plucking, striking, and bowing.

Plucking

Suppose a string is plucked with a sharp object at its midpoint. In the process, the string is made to assume an initial shape like the one pictured below. In order to determine which frequencies are present in the resulting vibration, we must determine which harmonics, when superimposed one upon the other, will add together to give a shape like the initial shape of the string. If the initial shape of the plucked string is compared with the waveforms of the first few harmonics, an educated

Initial shape of string
plucked at its midpoint

guess can be made about which harmonics will be present. Try it. First, note that the shape of the initial string configuration and the shape of the first harmonic are very similar. We might conclude that the relative intensity of the first harmonic is large. Next, note that the second harmonic has a node right where the amplitude of the string is initially a maximum. In fact, all even ($n = 2, 4, 6, \ldots$) harmonics have a node exactly where the initial shape of the string has its maximum amplitude. These harmonics, we might suspect, will be significantly depressed.

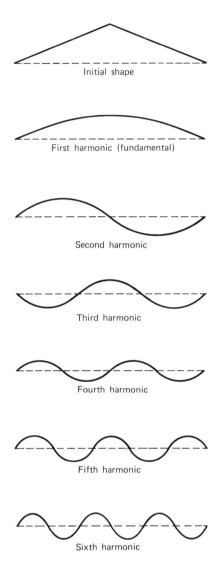

Initial shape

First harmonic (fundamental)

Second harmonic

Third harmonic

Fourth harmonic

Fifth harmonic

Sixth harmonic

Since the third and fifth harmonics have a maxima at the midpoint, they will be encouraged by plucking a string at its midpoint.

These suspicions are, in fact, valid. A string plucked at the midpoint has only odd harmonics present. The next set of figures show the superposition of (*a*) the fundamental and the third harmonic and (*b*) the fundamental, third, and fifth harmonics. These figures illustrate that the initial shape of the string is being approached by the superposition of the first three odd harmonics. If additional odd harmonics are included in the right proportions, the initial angular shape of the string can be even more closely approximated. The exact harmonic composition of a

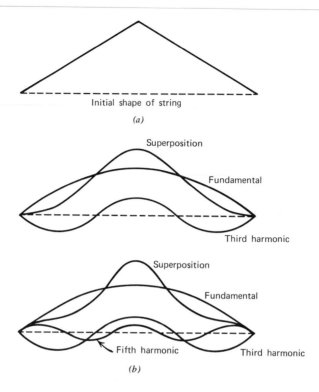

Initial shape of string

(a)

(b)

string plucked at its midpoint is shown in the sound spectrum below. Because the relative intensities diminish so rapidly, a table giving their precise values is shown beside the spectrum. (In this table, the minus signs indicate that f_3 and f_7 are 180° out of phase with f_1; f_1 and f_5 are in phase.) The fundamental is obviously the predominant frequency

Harmonic	Relative Intensity
1	1
2	0
3	$-\frac{1}{9}$
4	0
5	$\frac{1}{25}$
6	0
7	$-\frac{1}{49}$
8	0
9	$\frac{1}{81}$
10	0

Sound spectrum for a string plucked at the midpoint

present. Furthermore, the sound from a string plucked at its midpoint is characterized by the absence of even harmonics.

We can generalize the approach employed above to a string plucked at any point along its length. For example, suppose the string is plucked a distance $L/5$ from one end. The initial shape has a maximum amplitude right where the fifth, tenth, fifteenth, etc. harmonics have a node; hence these harmonics will be absent. The sound spectrum and table of the values graphed are given below for the first 10 harmonics.

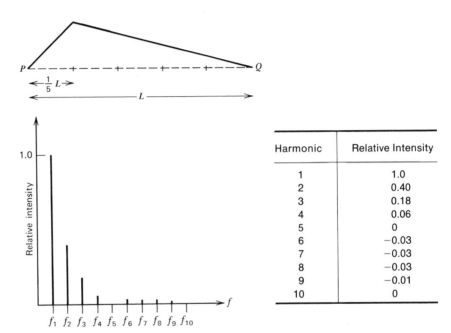

Harmonic	Relative Intensity
1	1.0
2	0.40
3	0.18
4	0.06
5	0
6	−0.03
7	−0.03
8	−0.03
9	−0.01
10	0

Sound spectrum of a string plucked a distance $\frac{1}{5}L$ from an end

Two conclusions can now be made. First, when a string is plucked, all harmonics having nodes at the pluck-point will be absent in the resulting vibration. Second, the closer to the end the string is plucked, the richer will be the harmonic content of the resulting vibration.

We have examined the harmonic structure of a string plucked with a sharp object. Therefore, the analysis is applicable to the banjo, plucked with a plectrum, and the harpsichord, plucked with a quill. Harps and guitars, on the other hand, are plucked with fingers. The initial shape of a finger-plucked string resembles the drawing below. The initial shape has a maximum amplitude along an increment of length ΔL. Any harmonic having a node along the increment ΔL will be suppressed with the result that the larger ΔL is, the more harmonics are suppressed. A

Initial shape of a string
plucked with a blunt object

plectrum-plucked string is richer in higher harmonics than a finger-plucked string.

The initial shape of a string plucked rapidly is different from that of a string plucked slowly. When plucked suddenly, the string's inertia causes portions to lag behind where they would be if the string was plucked more slowly. Since the string is more sharply peaked at the pluck point, those harmonics having nodes near the pluck-point will be stronger than they would be if the string was plucked slowly. Consequently, the harmonic content of a rapidly plucked string is richer than for a slowly plucked string.

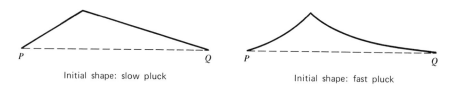

Initial shape: slow pluck Initial shape: fast pluck

A performer can exert considerable control over the harmonic content of the transmitted sound merely by the manner in which the vibration of the string is initiated.

Striking A string can be slowly and methodically plucked, but when it is struck, it is done so with impulsive vigor. In the plucking process the string is "slowly" deformed, thereby storing energy. At the instant the deformed string is released, the string's velocity is zero and its energy is entirely potential. By contrast, a string is "rapidly" struck by a massive object that is in contact with the string for approximately $\frac{1}{500}$ second. Thus, a stationary string is given an initial velocity and hence an initial kinetic energy. The resulting vibration of the struck string is more complicated than that of the plucked string.

The factors that influence the vibrational character of the struck string are manifold: the linear density of the string, the tension of the string, the mass of the hammer, the impact speed of the hammer, the contact time between hammer and string, the hammer dimension in contact with the string, the rigidity of the hammer, and the position of impact along the string. The design of a striking mechanism must be based

upon a physical analysis of these factors. For example, when a hammer strikes a string, a disturbance is sent propagating along the string in both directions. If a maximum amount of energy is to be transferred from the hammer to the string, the striking mechanism must be designed so that the hammer breaks contact with the string before the wave, reflected from the further end support of the string, returns to the impact point.

There are similarities between the struck string and the plucked string. A string struck in the middle creates an antinode where all even harmonics have a node; therefore, there will be no even harmonics. Likewise, the harmonic content of the vibration increases as the impact point approaches the end support of the string. Unlike the resulting sound of a plucked string, however, the relative intensity of the fundamental can be enhanced even when the string is struck near its end support merely by varying the mass of the hammer. In addition, if the hammer makes contact with the string over an increment of length ΔL, all those harmonics having nodes falling in that increment will be absent in the resulting vibration.

Bowing The nature of the excitation mechanism changes significantly when a bow comes in contact with a string. Plucking and striking put energy into the string impulsively. We shall think of these as momentary excitation mechanisms. Bowing, on the other hand, puts energy into the string on a continuous basis so that bowing is a sustained excitation mechanism. As a consequence of this continuity, the performer has much more control, and can influence the resultant sound more effectively.

We shall begin our analysis by comparing the shapes of the plucked and bowed strings. The changing shape of a plucked string can be understood in terms of the superposition of two waves traveling in opposite directions along the string. The *initial* shape of a string plucked at its midpoint (solid line) and the two waves (dotted line) are shown in part (i) of the figure below. An instant ($\frac{1}{20}$ period) after the string is released, the two waves are moving in opposite directions, their superposition determining the string's shape—see part (ii). Successive increments of time—each increment equaling $\frac{1}{20}$ period—are shown through the first quarter of a cycle—see parts (iii) to (vi).

The string's motion is somewhat surprising. Examine the midpoint (a). If you follow its motion through the first quarter cycle, you might be able to deduce the motion of the midpoint throughout the rest of its period. Check your reasoning by examining the graph below. Now

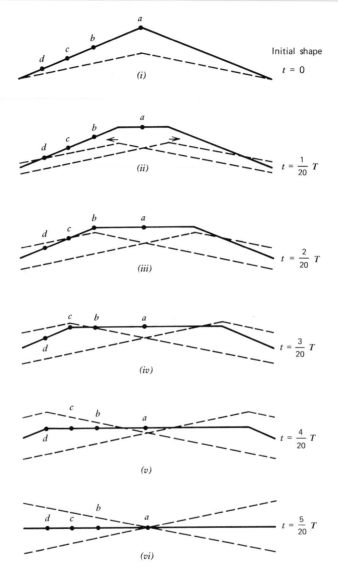

Shape of string, plucked at its midpoint, through $\frac{1}{4}$ cycle

consider point (c), a distance $0.2L$ from an end. How does it move throughout the cycle? Its motion along with the motions of points (b) and (d) are also graphed on page 107. The midpoint (a) is moving constantly up or down. Point (c) however, spends most of each period not moving at all. The graphs show that the closer a point is to the end of the string, the less time it spends moving. This is reasonable since the end points are always stationary.

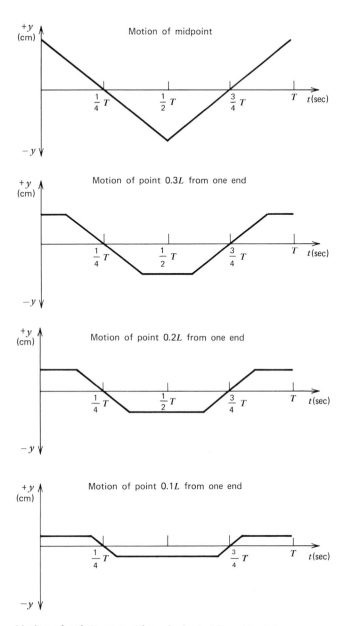

Motion of points on a string plucked at its midpoint

Also given are graphs representing the motion of a string plucked near an end. The motions of three points along the string are also shown.

From these two examples we can conclude that most points along a plucked string spend only a small fraction of each cycle in motion. Also,

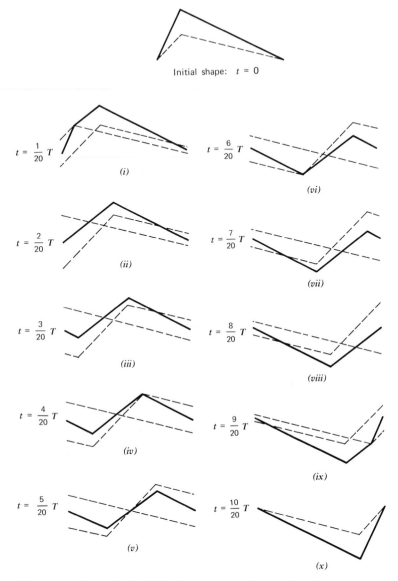

Initial shape: $t = 0$

$t = \dfrac{1}{20} T$ (i) $t = \dfrac{6}{20} T$ (vi)

$t = \dfrac{2}{20} T$ (ii) $t = \dfrac{7}{20} T$ (vii)

$t = \dfrac{3}{20} T$ (iii) $t = \dfrac{8}{20} T$ (viii)

$t = \dfrac{4}{20} T$ (iv) $t = \dfrac{9}{20} T$ (ix)

$t = \dfrac{5}{20} T$ (v) $t = \dfrac{10}{20} T$ (x)

Shape of a string, plucked 0.2 L from an end, through $\frac{1}{2}$ cycle

the shape of a plucked string throughout a cycle can be represented by three straight lines.

The motion and shape of a bowed string is initiated by the interaction between a bow and a string. When the bow is placed on the string and drawn steadily to one side, the string alternates between sticking to the bow and sliding against it. During the sticking portion the string moves with the bow. When the string slips, it does so in a direction opposite to the

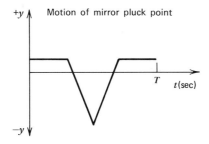

Motion of points on a string plucked 0.2 L from one end

direction of the bow's motion. This alternation between sticking and sliding is extremely regular and the basis for this regularity has been understood only recently. To the casual observer, the shape of the bowed string appears the same as the shape of a plucked string; however, looks are deceiving. If we study the bowed string with photographic techniques so that the vibrating string can be observed in slow motion, we find that the string has an angular shape formed by two straight-line segments. The junction point of the two line segments propagates along the string and is reflected at the end supports.

The salient features of the bowed string's motion can be represented in a diagram. Point *P*, the discontinuity, races around a parabolic path (dotted line) once during each cycle. For the open A-string on a violin,

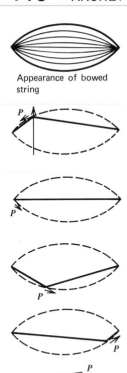

Appearance of bowed string

The shape of a bowed string through one cycle for the case of up bowing

$T = \frac{1}{440}$ second; therefore, the discontinuity makes 440 round trips every second.

The discontinuity passes under the bow twice—once while moving to the right and once while moving to the left—during each round trip. Each time the discontinuity passes under the bow the status quo is disturbed: if the string is sticking to the bow, it will be knocked loose by the passing discontinuity. If, on the other hand, the string is slipping along the bow, the passing discontinuity enables the bow to once again grab hold of the string. The revolving discontinuity is the clockwork mechanism that provides the precise alternation (referred to earlier) between sticking and sliding.

The motion of a bowed string is different from that of a plucked string. Likewise, the points along a bowed string move differently than they would if the string was plucked. The points along a plucked string spend only a fraction of each period in motion. This is not true for a bowed string. Let Q be a point on the string that is a distance D from the bridge. If the length of the string is L, the point Q divides the string into two segments having lengths D and $L - D$. For a string bowed in the "up-bow" direction, the time point Q spends slipping (falling) is proportional to D. The time point Q spends sticking (rising) is proportional to $L - D$. If $T_F =$ the time spent falling (slipping) and $T_R =$ the time spent rising (sticking), we can write

$$\frac{T_F}{T_R} = \frac{D}{L - D}$$

or

$$T_F = \left(\frac{D}{L - D}\right) T_R$$

If

$$D = \frac{1}{20} L = 0.05L, \quad \text{then} \quad \frac{0.05L}{L - 0.05L} T_R = \frac{0.05}{0.95} T_R = 0.053 T_R.$$

However, since $T_R + T_F = T$, we can write

$$T_R + 0.053 T_R = T$$
$$1.053 T_R = T$$
$$T_R = 0.95 T$$

This result indicates that 95 percent of each period the string is rising. The other 5 percent of the period it is falling.

These results are shown below in diagrammatic form. Three points Q, R, and S are identified on the string. The motion of each point is graphed separately. From both the results above and the succeeding graphs it is apparent that all points along a bowed string are moving at all times.

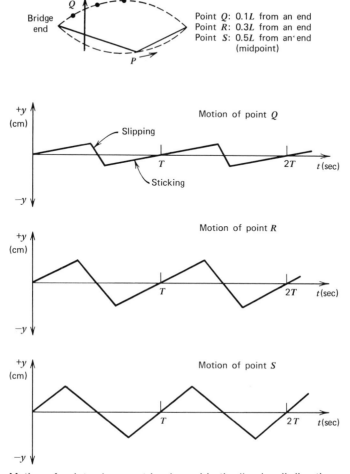

Motion of points along a string bowed in the "up-bow" direction

A string bowed in the "up-bow" directions initiates a discontinuity that revolves around a parabolic path in a counterclockwise sense. When the string is bowed in the "down-bow" direction, the discontinuity revolves in a clockwise sense. As the succeeding figures show, the ratio of times spent sticking and slipping is independent of the bowing direction.

The harmonics present in a string vibration depends on its shape. Since the shape of a bowed string is different from that of a plucked string, we would expect a difference in their timbre. The triangular form of the bowed string makes possible a sound spectrum that is very rich in harmonics. Both even and odd harmonics are present and their intensities decrease much more slowly than do the harmonic intensities of a plucked string. For the bowed string, the intensity of the nth harmonic is

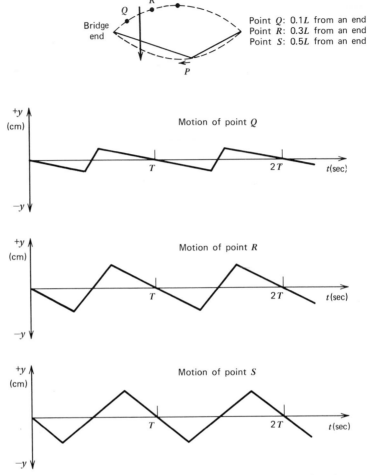

Motion of points along a string bowed in the "down-bow" direction

$1/n$ times the intensity of the fundamental. For the plucked string, on the other hand, the intensities of harmonics decrease in proportion to $1/n^2$. The rich harmonic content of the bowed string is of utmost significance in the production of sound in the string instruments.

In summary, when the bow is placed on the string and drawn steadily across it in the "up-bow" direction, the string alternately sticks to and slips against the bow. There is an initial period of instability before the standing waves get established on the string and the precise alteration begins. The bowing process is represented on page 113.

The shapes and motions of the bowed string and plucked string are different. For this reason, the quality of sound of a string excited by these two techniques is quite different.

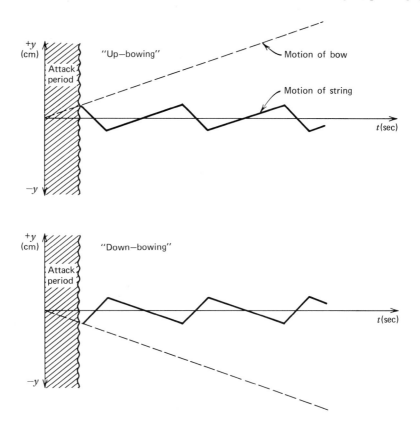

THE GENERATION OF COMPLEX VIBRATIONS IN AN AIR COLUMN

Anyone who has ever blown across the top of a soda bottle has probably been able to excite the enclosed air into vibration. If you can recall such an experience, you may remember that the instant you stopped blowing, the bottle "stopped singing." This suggests that the enclosed air vibrates only so long as you help it to do so. When the outside assistance ceases, the air vibrations decay immediately. In this regard, an air column behaves differently than a stretched string: a string's vibrations decay slowly while those of an air column decay immediately. Impulsive excitations, therefore, will not suffice for the air column; rather, energy must be supplied continuously if oscillations are to be sustained.

Edge tones

Imagine a chamber containing air at a pressure higher than that of the outside air. If a narrow slit is opened in a wall of the chamber, a jet of air will stream from the slit into the outside air. Since the outside air is relatively still, the edges of the jet stream will be slowed by the ambient air. In fact, the edges will be peeled back and there will be formed small whirlpoollike vortices. The sideways motion of the air in these vortices

exerts a sideways (or transverse) influence upon the faster-moving interior of the jet stream with the result that the stream will follow a wavy line. The formation of these vortices also disturbs the surrounding air.

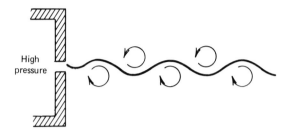

The rate at which these vortices are formed depends on the velocity of the jet stream and the dimensions of the slit. Under the proper conditions of velocity and slit dimensions, the periodic disturbance caused by the formation of vortices can become an audible sound source. Examples of this phenomenon are numerous: a person whistling, the whistling of the wind through a crack around a door or through a slightly opened car window, and the spooky sound of the wind around the eaves of a house or through the rigging of a ship.

The sound resulting from a jet stream of this sort is rather weak. The intensity of the sound can be increased by the addition of a wedge situated so that its edge bisects the jet stream. The vortices on each side of the jet stream move along opposite sides of the wedge but, in the process, the passing vortices exert a periodic force on the wedge

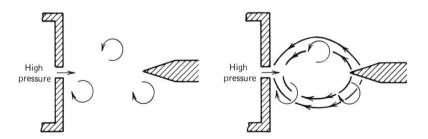

transverse to the jet stream. This periodic force initiates a chain of events: the wedge is set into vibratory motion, the vibrating wedge disturbs the contiguous air, and this periodic disturbance is propagated back to the jet stream. The effect of this periodic disturbance on the jet stream is to encourage the formation of larger vortices, which results in a more intense sound.

The edge-tone mechanism is an excellent example of *feedback*. We start with

1. A jet stream, inherently unstable, giving rise to vortices, which

2. Interact with the wedge, exerting upon it a periodic force.

Next, the wedge

1. Disturbs the surrounding air, which

2. Enhances the inherent instability of the jet stream and leads to

3. The formation of larger vortices, which results in

4. The amplification of the resulting sound.

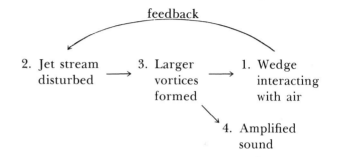

The feedback step is indicated in the flow diagram above. The closed path $1 \rightarrow 2 \rightarrow 3 \rightarrow 1$ is often called a feedback loop. As we shall soon see, feedback is crucial in the production of sound by the wind instruments.

With the addition of the wedge, the frequency of the edge tone depends on two factors: the velocity of the jet stream and the distance between the orifice and the wedge. The frequency increases as the velocity of the jet stream increases and decreases as the orifice-to-wedge distance increases. With some caution, the frequency can be expressed by the following proportionality:

$$f \propto \frac{v}{h}$$

where h is the orifice-to-wedge distance and v is the velocity of the air stream.

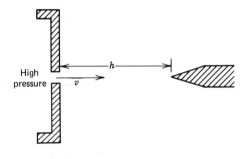

Let's see why caution is necessary. The frequency of the edge tone is determined by the rate at which vortices pass the edge. Suppose the jet stream velocity is fixed and the orifice-to-wedge distance is varied. At some separation, we can envision a vortex arriving at the wedge at the very same time a vortex is being formed at the orifice [see figure (a) below]. This condition defines the minimum separation for the fixed velocity that will give rise to an edge tone of frequency f_1. As the wedge-to-orifice distance is increased from the distance h_1, the frequency decreases to a new value f_2, that is, $f_2 < f_1$. [See figure (b) below.] At this increased separation, h_2 the frequency is likely to be unstable, jumping discontinuously back and forth between f_1 and f_2. The reason for this frequency instability can be understood by examining figures (a) to (c). At the increased separation h_2, the lower frequency f_2 is heard when the number of vortices between the orifice and the wedge has not changed in going

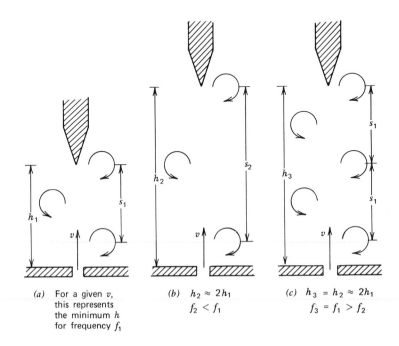

(a) For a given v, this represents the minimum h for frequency f_1

(b) $h_2 \approx 2h_1$
$f_2 < f_1$

(c) $h_3 = h_2 \approx 2h_1$
$f_3 = f_1 > f_2$

from the separation h_1 to h_2— see (a) and (b). However, this requires that the separation between vortices increase from s_1 to s_2. From the vortices point of view, this is stretching things a bit far; in fact, the vortices may rearrange themselves so that they have the smaller separation s_1 [figure (c)] in which case the rate at which the vortices pass the wedge is the same as it was with the smaller separation h_1 and a sound of the same frequency f_1 is produced. Other instabilities of this sort exist, which are well known to the flutist.

As the velocity of the jet stream increases, the frequency increases because the rate at which vortices pass by (and interact with) the wedge increases. If *both* v and h are increased together, the frequency can remain unchanged.

Our objective is now in hand. An air column can vibrate with many frequencies. Those frequencies actually present depend on how the air column is excited into vibration. The edge tone is one way of energizing an air column. Consider an edge-tone system (orifice and wedge) that is part of a cylindrical tube. The vortices on the "inner" side of the wedge provide periodic disturbances to the air column that set it into oscillation. By varying the velocity and orifice-wedge distance—a flutist rolls the flute toward or away from the lip orifice—different frequencies of the air column can be excited and the timbre influenced.

Open
or
closed

The air column exercises a controlling influence on the edge tone through the feedback phenomenon. When the vortices enter the air column, a standing wave is established in the air column, typically at the fundamental frequency. Air will thus be driven in and out of the opening between the orifice and the wedge at the frequency of oscillation of the air column. The jet stream emanating from the orifice is thus successively pulled into and pushed out of the opening by the action of the oscillating air column. At this point, the air column, through feedback, is completely dominating the edge tone and enhancing its stability. If the velocity of the jet stream is increased slightly, the frequency of the edge tone would increase if it could. But it cannot because the air column is in control. If the velocity of the jet stream is significantly increased, as in overblowing, the frequency will jump to the frequency of the next higher harmonic of the air column.

We see that the edge tone initiates the vibrations of the air column, but that the air column quickly controls the edge tone. The resulting sound is therefore the resonance frequencies of the air column.

Reed Tones Whereas the details of the edge tone and reed tone differ considerably, the relation between the air column and both the reed tone and edge tone are similar; namely, once the reed tone excites the air column into oscillation, the air column dominates the reed tone. Thus, to understand the reed instruments, the feedback concept is once again required.

A reed is a thin strip of metal, cane, or plastic. Metal reeds, designed to operate at one frequency only, are found in harmonicas and some pipe organs. The cane and plastic reeds, on the other hand, are softer and more flexible. They can be forced to vibrate at a number of frequencies. We shall only be concerned with the cane-type reed.

The cane reed is clamped at one end to a carefully designed mouthpiece that provides clearance for the free end to vibrate. A schematic diagram of a reed system is shown below. There is a space between the free end of the reed and the mouthpiece through which air can freely pass. When the pressure in the cavity is greater than the pressure inside the mouthpiece, air streams between the reed and the mouthpiece. The air pressure in this air stream (sometimes referred to as *dynamic* air pressure because the air is moving) is less than the static air pressure in the cavity; this forces the reed into contact with the mouthpiece thereby cutting off the air flow. With the air stream terminated the pressure on each side of the reed becomes nearly equal and the elastic properties of the reed snap it back to its original position. The cycle is repetitive with the reed opening and closing against the mouthpiece.

Each time the reed snaps back from the mouthpiece, a new burst of air passes through the opening. When the mouthpiece is fastened to a tube, the burst of air sets the air column into oscillation. In fact, the burst of air initiates a compression that propagates along the air column. At the end of the tube opposite the mouthpiece, the compression is reflected and a standing wave is established in the tube. Since the reed and mouthpiece behave like a closed end, the fundamental mode has a pressure antinode at the mouthpiece. For half of the fundamental's period the pressure in the mouthpiece is high and for the other half it is low. This variation of the air column pressure in the mouthpiece effectively controls the vibration of the reed. (One can confirm this by blowing the mouthpiece without the

tube and noting that it has a pitch quite different than the pitch perceived when the air column is in control.)

The pressure in the cavity (or player's mouth) is always higher than the highest pressure in the mouthpiece. Thus, when there is a pressure maximum in the mouthpiece and the reed opening is also a maximum, air enters the mouthpiece to resupply energy that has been lost through the production of sound. In this way, a standing wave of constant amplitude is sustained in the tube.

THE GENERATION OF COMPLEX VIBRATIONS IN A MEMBRANE

Like strings and air columns, a membrane can vibrate with all its overtone frequencies simultaneously. The relative proportions of the overtones actually present depend on how the membrane is excited. If a membrane is struck at its center, the fundamental is the principal frequency excited. Since f_2, f_3, f_5, and f_6 (see previous section) have a vibrational node at the center, these overtones will be absent.

To excite the largest number of vibrational modes, the membrane should be struck between the center and the rim, closer to the rim than to the center. Putting energy into the membrane in this fashion excites vibrational modes that decay slowly and give a lively sound.

STRUCTURAL INFLUENCES ON THE TRANSMITTED SOUND

Strings

Earlier it was pointed out that a string stretched between two sides of a window frame would be an anemic sound source. This is because the surface area of a string is so very small that the transfer of energy from the vibrating string to the contiguous air is a very inefficient process. (For the same reason, a toothpick makes a poor fan.) Considerable amounts of air must be moved if a big sound is to be achieved.

Vibrating strings need help if they are to move large quantities of air. The assistance comes from the structure on which the string is mounted. The vibrating motion of the string is transmitted by an intermediate structure such as a bridge to a resonator. In a piano, the resonator is the soundboard. In a violin, both the body of the instrument and the enclosed air space are resonators. In both cases, the resonator has a relatively large surface area that enables it to effectively transfer its own vibrational energy to the contiguous air.

The resonator, however, has its own vibrational characteristics. The vibrating string transfers a spectrum of harmonic frequencies to the resonator. The resonator responds to some of these frequencies more vigorously than to others. Consequently, some frequencies from the string are amplified by the resonator while others are attenuated.

The response of a resonator to a particular frequency may be determined by forcing it to vibrate at that frequency. If the resulting amplitude is small, the response is poor; if the amplitude is large, the response is good. The *response curve* shown below is obtained by forcing

the resonator to vibrate at many different frequencies and determining the response, R, at each frequency. The response curve illustrated shows that the response is relatively strong at f_2, but weak at f_1. We can define the response, R, more specifically. If the resonator is forced into oscillatory motion by a reference signal of intensity I_{ref} and, if the output signal from the resonator has an intensity I, then R can be written

$$R = 10 \log \frac{I}{I_{ref}}$$

R is expressed in db. We shall see later that response curves can be very complicated.

We can now understand how the response of the resonator influences the transmitted sound. Suppose a string is vibrating such that its harmonic content is given by the audio spectrum on page 121. These vibrational frequencies are coupled to a resonator whose response curve has the form indicated. It is clear that the response of the resonator is greater for f_5 than for either f_1 or f_{10}. The transmitted sound is the product of both these graphs.

From these figures, it is apparent that the strength of the transmitted sound resides in the second, third, and fourth harmonics. Unless the string vibration is sustained, however, the quality of the transmitted sound will change as the sound intensity diminishes. Since the resonator is extracting energy more rapidly at frequencies f_2, f_3, and f_4, these frequencies will decay more quickly. The initial sound transmitted from the resonator will be characterized by the dominance of f_2, f_3, and f_4 while the lingering sound will be characterized by their absence.

Tubes A bugle is a cylindrical tube flared at one end. The flare, called a bell, affects both the frequencies of the standing waves inside the instrument and the frequencies of the radiated sound.

Let us consider a cylindrical tube 137 cm in length with a radius of 0.6 cm. (These dimensions are approximately those of a trumpet.) With

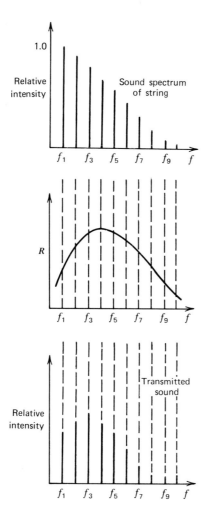

one end of the tube closed, the air in the tube will resonate at frequencies of f_1, $3f_1$, $5f_1$, etc., where $f_1 = 63$ Hz. [See the figure (a) on page 122.] Let us remind ourselves of two points.

1. These resonant frequencies are precisely those frequencies of the standing waves that can be accommodated by the tube.
2. Standing waves are only possible if waves are reflected by the open end.

However, if the tube is to be musically useful, acoustic energy must be radiated from the open end.

What determines whether a wave will be reflected by the open end or radiated from it? The answer is that it depends on the relative sizes of

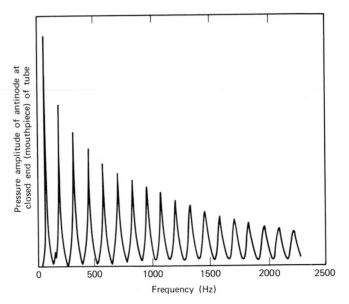

(a) Resonances of a trumpetlike tube

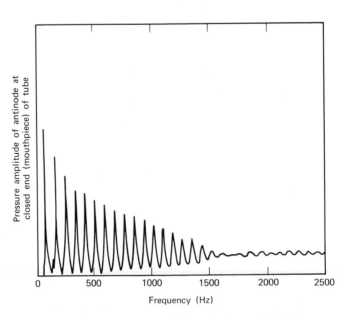

(b) Resonances after adding a bell to the trumpetlike tube

Parts (a) and (b) from *Fundamentals of Musical Acoustics* by Arthur H. Benade. Copyright © 1976 by Oxford University Press, Inc. Reprinted by permission.

the wavelength of the standing wave and the diameter of the tube. If the wavelength is much longer than the diameter of the tube, the acoustic energy is virtually imprisoned in the tube and little, if any, sound emanates from it. For the trumpetlike tube discussed above, essentially no sound escapes from the tube for frequencies below 612 Hz—the ninth harmonic.

Suppose we now add a bell to our tube. In a small diameter tube [i.e., diameter $\ll \lambda$], a wave cannot easily make the transition between the confinement of the tube and the boundless environment outside the tube. The bell facilitates this transition; however, if the bell flares too abruptly, the sound wave is unaffected by the bell's presence. If the bell is to be effective in assisting the escape of acoustic energy from the tube, the diameter of the tube must increase slowly over a tube length equal to the wavelength of sound. Such a wavelength will then "see" a larger opening and its escape will be enhanced.

The bell not only affects the radiation efficiency of the tube, it also alters the resonance frequencies *inside* the tube itself. Figure (*b*) on page 122 shows the new resonances inside the tube that result from the addition of the bell. [Figure (*b*) should be compared with (*a*).] With the bell added to the tube, resonance peaks end at about 1500 Hz. For higher frequencies (small wavelength) so little of the sound wave is reflected from the open end that standing waves cannot be established inside the tube. Furthermore, the addition of the bell squeezes the resonance peaks closer together. We shall discuss this in greater detail when we meet the brass instruments.

The bell radiates more effectively at high frequencies. Consequently, the sound spectrum of the complex tone inside the tube is different than the sound spectrum of the transmitted sound. This is illustrated on page 124.

In the same way that the sound transmitted by a string is modified by its immediate environment, the sound from a tube is also modified. When we discuss the brass instruments in more detail, we shall learn that the mouthpiece also influences the standing wave frequencies and thereby exerts an influence on the transmitted sound.

Membranes

A membrane supported on a ring would be unable to radiate significant amounts of acoustic energy. For such a membrane, a compression is created on one side at the same instant a rarefaction is formed on the other side. The two waves, having opposite phases, would effectively cancel each other out. When the membrane is stretched over a kettle, however, as with timpani, the sound wave from the underside cannot interact with the sound wave from the upper side. The radiating ability of the membrane is enhanced by the enclosed cavity over which the membrane is mounted.

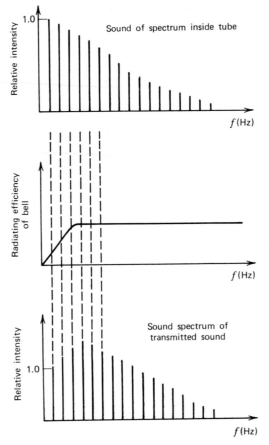

Influence of bell on the transmitted sound

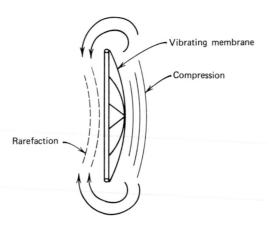

Strings, air columns, and membranes are principal sources of musical sound. The detailed technique used to excite these sources into oscillation as well as the immediate structural environment determine the character of the transmitted sound.

REFERENCES

1. Schelleng, John, "The Physics of the Bowed String," *Scientific American* **230**, 87 (January 1974).

In this article some aspects of the bowed string are discussed in greater detail.

2. Powell, Alan, "On The Edgetone," *Journal of the Acoustical Society of America* **33**, 395 (1961).

A classic paper concerning the edge tone. This paper presents experimental data and an analysis of the data. Somewhat difficult to understand.

3. Benade, Arthur H., "On Woodwind Instrument Bores," *Journal of the Acoustical Society of America* **31**, 137 (1959).

This article discusses the shapes of horns that are musically useful.

4. Backus, John, "Vibrations of the Reed and the Air Column in the Clarinet," *Journal of the Acoustical Society of America* **33**, 806 (1961).

The nature of reed vibrations are studied experimentally.

For Further Thought

1. A stretched string is 60 cm long. It is plucked with a plectrum at a point 20 cm from one end.
 (a) What harmonics would you expect to be in the resulting vibration?
 (b) Indicate what the sound spectrum might look like.

2. A flexible string of length 99 cm and mass 0.001 kg is stretched to a tension of 110 newtons.
 (a) What is the linear density of the string?
 (b) If the string is plucked with a sharp object at a point 24.25 cm from one end, what are the four smallest harmonic frequencies in the resulting vibration?

3. The figure below shows the initial shape of a plucked string. Which harmonics would be absent from the complex vibration?

4. The figures on page 107 show the motions of specific points along a string when it is plucked at its midpoint. From the appropriate figure determine the fraction of each cycle that a point $0.1L$ from one end of the string is stationary.

5. On the basis of our analysis, why is the harmonic content of a bowed string different from that of a plucked string?

6. Consider a stretched string 30 cm long. If the string is up-bowed, what time does a point 6 cm from one end spend rising? Express your answer as a percentage of the period of oscillation.

7. Discuss the feedback phenomenon in the edge tone.

8. The frequency of an edge tone depends on the velocity of the jet stream emanating from the orifice. How does the frequency change as the jet stream velocity increases! Why?

9. With the jet stream velocity constant, some orifice-to-wedge distances lead to instabilities in the frequency. How do you account for these instabilities?

10. In a homemade instrument, strings are mounted on a wood base. The response characteristics of the base are evaluated by driving the base at various frequencies with a reference signal having an intensity of $I_{ref} = 10^{-6}$ watt/m^2. The data collected is given below.

Frequency (Hz)	Intensity (watt/m²)
100	1×10^{-4}
200	4×10^{-4}
300	8×10^{-4}
400	2×10^{-3}
500	3×10^{-4}
600	2×10^{-3}
700	9×10^{-4}
800	5×10^{-4}
900	2×10^{-4}
1000	1×10^{-4}

From this data, generate a response curve for this wood base. (You might find it useful to review logarithms in Chapter 3. Also, see question 1 in Chapter 4.)

11. From the response curve obtained in the previous problem, suggest how the transmitted sound would be shaped by the resonator if a string mounted upon it was tuned to oscillate with a fundamental frequency of 100 Hz and if the string was
 (a) Plucked at the midpoint.
 (b) Plucked at a point $\frac{1}{4}L$ from one end.

12. A tube has a length such that the enclosed air can oscillate with a frequency of 262 Hz. The temperature of the air is 22°C. Two bell sections are available that can be added to the tube: one bell section is tapered over a length of 16 cm; the other bell section is tapered over a length 48 cm. If the bell of each section has the same diameter, which bell would radiate sound of frequency 262 Hz more efficiently?

10 The Perception of Complex Tones

This is a short chapter.

A complex wave incident upon the eardrum sets it into vibration with a waveform that has the same shape as that of the incident sound wave. The waveform is transmitted intact to the inner ear. Neither the eardrum nor the middle ear know that they are transmitting a complex vibration consisting of the superposition of many simple harmonic oscillations. When the signal reaches the inner ear, however, the complex tone is separated into its individual frequency components. Each pure-tone component activates a different set of neurons along the basilar membrane.

A trombone playing a concert C_3 produces a complex tone with harmonics having frequencies of 131, 262, and 393 Hz. With all these frequencies present in the inner ear, a puzzling question can be asked: how is it that we are able to associate a definite pitch to a complex tone? Fundamental tracking is apparently the mechanism that enables us to assign a definite pitch to a complex tone. The figure below shows a C_3 note (solid) and higher harmonics (open notes). If consecutive harmonics are considered a pair at a time, the repetition rate of their superposition curve is equal to the frequency of the fundamental. It is believed that our auditory system uses this commonly occurring repetition rate to identify the pitch of a complex tone. This explanation is

supported by the fact that even in cases where the fundamental of a complex tone is missing entirely, we are able to assign a pitch (that of the fundamental) to a complex tone. For example, in poor quality speaker systems low frequencies may not get through. A complex tone having a fundamental frequency of 100 Hz coming into such a speaker system may have only its higher harmonics of frequencies 200 and 300 Hz transmitted; yet, we identify the pitch as 100 Hz!

A second puzzling question can be raised: how is it that two complex tones having the same pitch and the same loudness can be distinguished from each other? In other words, how can we distinguish one musical instrument from another? A complete answer cannot be given to this question. A partial answer, however, can be attempted. An important part of the sound associated with every musical instrument is the attack and decay of a tonal sound. The slow buildup of sound in the flute is in marked contrast to the percussive initiation of a piano sound.

The significance of the attack and the decay in identifying the sound of a musical instrument can be demonstrated by a simple experiment. If the tape recording of a familiar instrument is played backwards, an entirely new auditory experience results. Even though the timbre is not affected by playing the tape in the reverse direction, tonal attacks become decays and decays become attacks. The result is that the instrument sounds unfamiliar.

Questions pertaining to the perception of complex tones are a topic of current research. They are interesting, but difficult questions. Much work must be done before definitive answers can be given.

The Whole Orchestra

When the maestro steps to the podium and lifts the baton, he commands strings of varying lengths and densities, tubes of varying lengths and radii, as well as membranes and other percussive devices. The strings are mounted on resonators of diverse shapes and different sizes. They are excited into vibration by bowing, striking, and plucking. The air columns are excited into vibration by edge tones, single and double reed tones, and buzzing lips. Mouthpieces, bells, and mutes of various shapes and styles add diversity to the sounds. These are the raw materials of the orchestra.

There are more than 100 performers in the modern orchestra. The typical orchestra of today is much larger than those of earlier periods. Orchestras during Bach's professional life had approximately 20 members while 40-odd players characterized the orchestras of Mozart's time. The modern orchestra is not only bigger, it is better. Contemporary musicians, unlike their predecessors, are able to give full time to rehearsal and performance. Consequently, modern orchestras achieve a level of precision unattained in earlier times.

When the raw materials of musical sound are placed in the hands of performers, we have the modern orchestra with its diversity of tonal resources, range of dynamics, and varieties of tonal colors. It offers a memorable experience to the music lover.

Performers in an orchestra are divided into four sections. The string section is the largest. The violin, viola, cello, and double bass make up this "heart of the orchestra." The woodwind section consists of flutes, oboes, clarinets, and bassoons along with the piccolo, the English horn, the bass clarinet, and the double bassoon. The brass section consists of

French horns, trumpets, trombones, and the tuba. Finally, the smallest section of the orchestra is the percussion section. In addition to the timpani, there are bass and snare drums, cymbals, triangle, xylophone, and other instruments in this orchestral section. We shall not consider all of these instruments. Rather, we shall consider a few representative instruments in moderate detail and thereby establish a basis for understanding other musical instruments.

Orchestras perform in concert halls. The acoustical environment enters actively into the musical process. Composers have been influenced by the acoustical characteristics of those structures accessible to them. Artists respond to the acoustical setting and modify their performance accordingly. Listeners are treated either to an audio performance where the orchestra is balanced and the sounds blend together properly or they endure the converse.

We shall examine those factors that influence the acoustical character of a concert hall. These factors are both objective and subjective in nature. We shall also examine the symbiotic relationship between architecture and music.

An orchestra produces many complex tones of many different frequencies. Some of the tonal combinations strike the perceiver as more pleasant, others as less pleasant. A consonance is a musical interval deemed pleasing by the typical listener while a dissonance is judged unpleasing. These are elusive terms. "Pleasing" and "unpleasing" are obviously subjective experiences derived in part from cultural conditioning. A musical interval considered a consonance today may well have been considered unacceptable by composers and artists in an earlier period.

In spite of the subjective dimension of consonance and dissonance, there are objective considerations that can be clarified. We shall examine the objective aspects of consonance and dissonance as well as musical scales.

11 The Instruments of Musical Sound

In the previous chapters we introduced concepts that are necessary if we are to understand the origin of musical sound, its transmission, and its perception. Let us now bring these concepts together and employ them to understand selected musical instruments that in the present context can be considered as the origin of the sound we call music. If the selection is made judiciously, insight into many other instruments will be achieved in the process.

THE STRINGED INSTRUMENTS

The orchestral stringed instruments consist of the violin, the viola, the cello, and the double bass. These are principally bowed instruments; however, some passages call for other special effects. As it can be seen from the figure below, the violin has the widest range. Because of its smaller size, the violinist can play more complex rhythmic ideas and has greater agility in rapid passages. The viola, tuned a perfect fifth lower in pitch than the violin, is correspondingly larger. The cello is tuned one octave below the viola and its size dictates that it be played differently; namely, the instrument stands on the floor. Both the cello and the violin often carry the melody in the orchestra or fill in the harmony. The double bass provides the harmonic foundation for the string section. Frequently, the double bass is accompanied by the cello playing an octave higher. With this arrangement, the double bass assumes great carrying power and furnishes the basic support for the string section. (See question 1 below.)

The violin, viola, cello, and double bass each have four strings, a

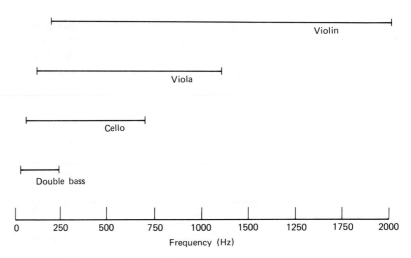

Frequency ranges of the orchestral stringed instruments

bridge, a top "plate," a bottom "plate," and sides connecting the two plates, thereby forming a volume of air. These components form a complex system. Each component has its individual natural frequencies of vibration as well as its individual damping characteristics. Yet each part is intimately coupled to another part in such a close relationship that each part sacrifices its individuality for the benefit of the composite whole.

The four strings of each of these instruments are stretched over a long, thin, smooth wooden structure called a fingerboard. The performer varies the length of the string (and, hence, the family of harmonic frequencies) by pressing down a finger of the left hand at a particular point along the fingerboard. This action creates a displacement node and leaves a shorter length of string free to vibrate.

The stringed instrument that has been studied the most intensively by both physicists and acousticians is the violin. I shall focus my comments on this remarkable instrument, but bear in mind that the insights we arrive at here can be extended to the other members of the stringed family.

The Violin The exact origin of the violin is unknown. Perhaps Andrea Amati, who died around 1580, invented this instrument. In any event, Amati is credited for having founded the Cremona school that brought the center of violin making to Northern Italy.

The close relationship between the technician and the artist is no where more apparent than in the development of the violin. The years between 1650 and 1750 are so intimately bound up between great violin

makers and great violinists-composers that it is difficult to determine whether violin music directed the builders or violin builders guided the musicians. The form given to the violin during these decades has remained the standard by which present-day fiddles are made.

(i) Its Structure

In simplest terms, a violin is a set of strings stretched over a wooden box. The violin strings can be metal, pig gut, or gut wound with either silver or aluminum wire. The box is hollow, delicately shaped, and is fragile. The strings are stretched from the tail piece—across the bridge and over the fingerboard—to the nut end. When the strings are properly tuned, the total tension of the four strings is about 50 pounds or 222.4 newtons. Consequently, a force of approximately 20 pounds is directed downward through the bridge against the fragile top plate of the box.

Some mechanical support is required to keep the arched top from being crushed under the 20 pound load. The necessary strength is provided in two ways. First, by a cylindrical post mounted under one leg of the bridge between the top and bottom plates of the box. It is called the sound post and is held between the two plates by friction. Second, a strip of wood mounted under the other leg of the bridge runs along the underside of the top plate beneath the string tuned to the lowest pitch. Because of its location—under the lowest-tuned string—this strip of spruce is called the bass bar. Both the sound post and the bass bar dramatically influence the acoustical character of the violin.

Near the outside edge of both the top and bottom plates is a shallow groove. Inlaid in this groove is the purfling that consists of two strips of black-dyed pearwood plus a strip of white poplar. Since the plates are thin to begin with, the groove makes the edges extremely thin. After years of playing the purfling tends to loosen somewhat making both plates more flexible. This may be a factor in the improved tone qualities of older instruments. Finally, a matching pair of delicately shaped "f-holes" are cut into the top plate on each side of the bridge. The long axis of the instrument divides both the top and bottom plates into two symmetric halves.

The structure of the violin body is one of the more remarkable stories in the history of music, even though the story is far from complete. Acoustical knowledge apparently played a very small role in the evolution of the violin; rather, it was the product of ingenious hunches, tedius trial and error, and the brilliant intuition of the artisan. How was it determined, for example, that:

—Maple was best for the bottom plates?
—The maple should be cut in December or January (when the sap flow stopped) and furthermore that the wood should be taken from the south side of the tree?

—The wood should be dried at least 10 years, but not kiln dried?

—Spruce was best for the top plate?

—The best spruce came from mountain slopes where the tree was exposed to the sun?

—The top (and bottom) plate should be thick in some places while thin in other places?

These questions, and others like them, have no definite answers. It is perhaps because of these mysteries that so many myths and so much folklore has surrounded the history and development of the violin.

The structures of the viola and cello are essentially the same as the violin. A diagram identifying the various parts of the violin is shown below and is followed by an exploded view of the viola.

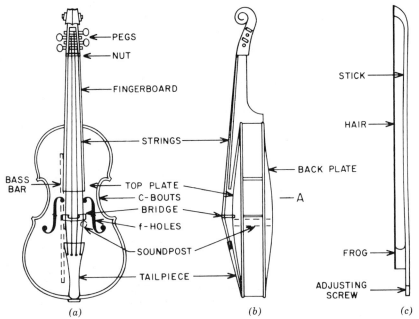

(Reprinted from *The Acoustical Foundations of Music* by John Backus. By permission of W. W. Norton & Co., Inc. Copyright © 1969 by W. W. Norton & Co., Inc.)

(ii) Its Acoustical Characteristics

When the bow is drawn across a string, it is set into vigorous vibration. As we learned in Chapter 9, the sound spectrum of the bowed string is rich in both even and odd harmonics. The *origin* of the sound that emanates from the violin is the vibrating string; however, the string alone would not even provide an anemic substitute for the familiar sound of this instrument.

The vibrations of the string are transmitted to the body of the instrument via the bridge and sound post. The string oscillation forces the bridge into a rocking vibration—the vibration taking place in the plane of the bridge itself. The rocking bridge forces the top plate into oscillation and, via the sound post, the bottom plate is also forced into oscillation. The body of the violin, therefore, becomes intimately involved in the sound-production process.

The body of the violin actively shapes the spectrum of the emitted sound. This is because the wooden structure has scores of frequencies at which it vibrates naturally. When the frequency of the string vibration equals one of the natural frequencies of the wooden structure, the response of the violin body—through the resonance phenomenon—is very large. The result is that the pitches corresponding to the resonant frequencies of the violin body are amplified to a greater extent than are other frequencies.

But there is still more. The volume of air defined by the top, bottom, and sides of the violin body also participates in the spectrum-shaping process. Just as a Coke bottle can be made "to sing" when the enclosed air is forced to oscillate at its resonant frequency, the air in the fiddle body can be excited by gently blowing over the "f-holes." The oscillation of the air space in the violin is communicated to the outside air via these "f-holes." When the frequency of a string vibration equals the resonant frequency of the enclosed air space, that frequency will be selectively amplified.

Wood is a complicated material, no two pieces are exactly alike. In addition, the body of the violin is a complex structure. It is reasonable to expect that the response curve of the violin will be complicated. This expectation is verified by the figure below. This curve was obtained by

(Carleen M. Hutchins and Francis L. Fielding "Acoustical Measurement of Violins" *Physics Today*, 1968.)

some clever experimentation. The violin was suspended by rubber bands attached to the four corners of the instrument. A device, called a transducer, which converts an electric signal into a mechanical signal, was attached to the violin. The violin is forced to vibrate at a frequency determined by the frequency of the electrical signal. With the amplitude of the driving frequency kept constant, the amplitude of the violin's vibrations can be determined by monitoring the sound radiated from it with a sound-level meter. At the resonant frequencies, the violin vibrates more vigorously as indicated by the peaks in the response curve. (For more details, see reference 1 at the end of this chapter.)

Two of the peaks in the response curve have names: the *main wood resonance* (MWR) and the *main air resonance* (MAR). The main air resonance is the lowest resonant frequency of the enclosed air space in the fiddle. The main wood resonance is the lowest resonant frequency of the fiddle body. The two peaks corresponding to the MAR and the MWR are among the first few peaks on the left side of the response curve illustrated. The exact frequencies at which the MAR and the MWR occur are very important and determine the quality of the violin.

The relationship between the MAR, MWR, and the quality of an instrument can be more clearly ascertained from a different kind of response curve called a *loudness curve*. To determine a loudness curve, a violin is bowed without vibrato so as to produce the loudest tone possible at each semitone interval throughout the frequency range of the instrument. The sound-intensity level is determined for each semitone with a sound-level meter and the results are plotted on a graph.

The loudness curves for two violins are shown below. The lowest frequency of the violin is the open G-string; therefore, this frequency defines the origin of the graph. The other open-string frequencies are identified on the graphs by vertical lines intersecting the frequency axis. From these loudness curves it is clear that some pitches are considerably more intense than others. The main air resonance is identified on each loudness curve by an open circle and the main wood resonance by a solid circle.

The frequencies of the MAR and the MWR for the two violins are clearly different. In loudness curve (*a*), the MAR and the MWR are seven semitones, or one perfect fifth, apart. Furthermore, the MWR is located just below (within a whole tone) the open A-string frequency of 440 Hz. The MAR, always lower in frequency than the MWR, is just below the frequency of the open D-string. In loudness curve (*b*), the MWR is considerably higher in frequency than the open A-string and more than a perfect fifth above the MAR (which is rather weak).

In both response curves (*a*) and (*b*) there is another resonance peak one octave below the main wood resonance. It is identified by an x. In acoustical literature this peak is called the "wood prime" peak. When the MWR is properly located (near 440 Hz), the "wood prime" resonance is

located near the open G-string frequency (near 220 Hz) and thereby reinforces the lower notes of the G-string.

The loudness curve (a) is for a good 1713 Stradivarius and (b) is for a poor 250-year-old instrument of unknown origin. The frequencies of the MAR and MWR determine whether the instrument is judged good or bad. In tests on many instruments it has been found that good instruments have their MAR and MWR a perfect fifth apart and their frequencies nearly coincide with the unstopped-string frequencies of the two middle strings. In response curve (b), there is too wide a frequency range between the MAR and the MWR within which the response is weak. The result is an instrument of poor quality, lacking power and expressive qualities.

By looking at the response curve and the loudness curves, we can make an observation. The resonance peaks are denser (closer together) at high frequencies than they are at low frequencies. This means that the high-frequency response of the instrument will be more uniform than the low-frequency response. At low frequencies, the resonant frequencies are more specific and their frequency location—as we have seen—is critical.

How does the violin maker proceed so that the resonant frequencies of the finished product will be properly located? What guides the craftsman during the construction process? To begin with, the indi-

vidual parts have their own natural frequencies. The traditional method of the violin craftsman has been to listen to the *tap tones* of the front and back plates. To do this, the maker holds the plate near one end between his thumb and forefinger and taps the plate at various places and listens carefully to the pitch of the resulting sound. These sounds are referred to as tap tones.

As the top and bottom plates are thinned in the final phases of their construction, both the quality and pitch of the tap tones change. The violin maker works to achieve a clear belllike tone that has the correct pitch. As early as 1840 it was reported by Savart that "a top of spruce and a back of maple tuned alike produced an instrument with a bad, weak tone." He actually dismantled a number of Stradivarius and Guarnerius instruments and determined their tap-tone frequencies. He found that the pitch of the back was always one semitone higher in frequency than the top. Furthermore, he found that the tap tone of the tops varied between 138.6 Hz (C_3^{\sharp}) and 146.8 Hz (D_3) while the back plates varied between 146.8 Hz (D_3) and 155.6 Hz (D_3). More recent studies have confirmed that the tap tones of the top and bottom plates cannot have the same frequency, nor can they be more than a whole tone apart in pitch if the resulting instrument is to be judged favorably. (See references at end of the chapter.)

The finished instrument also has a characteristic tap tone. The first strong low-frequency resonance (marked with an arrow) in the response curve shown earlier corresponds to the tap tone of the particular violin tested.

As the violin maker designs and cuts the f-holes, he is having a determining influence on the frequency of the MAR. The air within the fiddle body is a somewhat complicated example of a Helmholtz resonator. Imagine a vessel such as that illustrated left. The volume of the vessel is V, the length of the neck is l and the cross-sectional area of the neck is A. The resonant frequency of the air enclosed in such a vessel is given by

$$f = \frac{v}{2\pi} \sqrt{\frac{A}{lV}}$$

where v is the speed of sound in air.

The formula above indicates that f increases as A increases while f decreases as V increases. In our example of the violin, the f-holes are the mouth of the vessel, l is the thickness of the top plate where the f-holes are cut, and V is the volume of the air enclosed in the violin body. The MAR therefore depends on the area of the f-hole openings and on the air volume in the violin. It should be pointed out that any attempt to substantially modify the frequency of the MAR by altering either A or V is not only risky, it is also impractical. In order to change the MAR

frequency by one whole tone, a 20 percent reduction in volume or a 59 percent increase in the f-hole area is required.

Finally, the violin maker can influence the resonant properties of the violin by the design of the bass bar and sound post. Small changes in the wood quality, the tightness of the construction, and the position of the sound post result in a greatly altered tone of the violin. So critical is the sound post that the French call it the soul of the instrument. If the sound post is removed entirely, the violin sounds something like a guitar.

The bass bar is also critical. Small changes in the height of the bass bar can shift not only the frequency of the tap tone, but also the shape of the tap-tone resonance. The removal of only a few shaves of wood causes noticeable effects. In one reported case, the successive removal of a mere 0.5 gram of wood from the bass bar modified the tap-tone resonance from its initial condition (which was poor), through an acceptable condition, and back to a poor condition.

Violin making is still an art. Experience and intuition are needed to guide the craftsman as the plates are thinned, the bass bar is shaped, or the sound post adjusted. Acoustical studies of the violin, however, are aiding the violin maker and are bringing us closer to an understanding of why some violins are judged fine and others are deemed poor.

Other Members of the Orchestral String Family

It has been established that in the best violins the MWR and MAR have frequencies within a whole tone of the two open middle string frequencies. This is not true of the viola and the cello. As a result, the lower notes of these instruments suffer in strength and quality.

Typically, the MWR and MAR of the viola and cello fall at frequencies three or four semitones higher with respect to the unstopped frequencies of the two middle strings than they do in the violin. These resonances are therefore too high to provide adequate support for frequencies in the lower range of these instruments. The reason for this in understandable. These difficulties emerge as a consequence of the compromises that are made between instrument size and playing convenience.

The best approach for understanding the inherent differences between the stringed instruments is through scaling theory. Let us consider the violin as our reference. The viola and cello are tuned a fifth and a twelfth, respectively, below the violin. This means that the longest wavelength of sound emitted by the viola is 1.5 times longer and that emitted by the cello is 3.0 times longer than the longest wavelength transmitted by the violin. (This is equivalent to saying that the lowest frequency of the violin is 1.5 and 3.0 times *higher* than the lowest frequencies of the viola and cello, respectively.) We could conclude that the strings of the viola should be 1.5 times longer than those of the violin while those of the cello should be 3.0 times longer. Also, we might

conclude that all the other dimensions of the viola and cello should be related to the dimensions of the violin by the factors 1.5 and 3.0, respectively. This is not the case.

The body dimensions of the viola are larger than those of the violin by the factor 1.17 and the cello is larger by the factor 2.13. These factors represent a significant departure from the scale factors 1.5 and 3.0. It is this departure that is responsible for shifting the MAR and MWR from those frequencies coinciding with those of the open middle strings. Recall, for example, that the frequency of the air resonance is given by:

$$f = \frac{v}{2\pi} \sqrt{\frac{A}{lV}}$$

Since the dimensions of the cello are approximately twice those of the violin, the air resonance will be approximately an octave lower ($\frac{1}{2}$ the frequency) than the air resonance of the violin. This puts the air resonance near D_3 rather than the desired frequency near G_2. The lower frequencies of the cello are therefore inadequately supported by air and wood resonances.

It is reasonable to assume that practical considerations as to how these instruments should be played guided the artisans in the design of the viola, cello, and double bass. A viola 1.5 times larger than the violin could not be played in the conventional manner. A cello 3.0 times larger than the fiddle would be as large as the present-day double bass and would have to be both played and tuned differently. (The strings in the present-day double bass are tuned a fourth apart—E_1, A_1, D_2, and G_2—to minimize the necessary finger spacings and the distances traversed by the hand.) The double bass is tuned one octave below the cello. From the scaling theory we can conclude that the double bass should be twice the theoretical size of the cello or six times larger than the violin. Such a double bass would be twice as large as its present size. (The double bass is 3.09 times larger than the violin.) Clearly an instrument of such a size would present practical difficulties in handling. The compromises made between scaling and playing convenience have resulted in shifting the MAR and MWR to higher frequencies leaving the lower frequencies of the viola and cello without the resonance support comparable to that of the violin. Loudness curves for the viola and cello are shown on page 144. Comparison of these curves with the analogous curve for the violin shows the critical differences.

The "Wolf" Tone The frequencies emanating from a stringed instrument are normally determined by the strings. There is the possibility, however, that the body of the instrument can challenge this exclusive right of the strings. When this happens, an agonizing sound, called a "wolf" tone, results.

Viola loudness curve

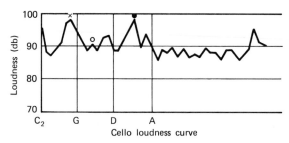

Cello loudness curve

This malady is most common with the cello, although both the viola and violin are occasionally afflicted.

The origin of the wolf tone is understood. It results from the interaction between the string vibration and a wood resonance. As we have seen, the body of a stringed instrument has many resonances (look again at the response curve of the violin). The main wood resonance is one of the most prominent of all the resonances and typically the MWR is involved in the generation of the wolf note.

If one of the heavier strings is commonly bowed at the frequency of the MWR, the wolf tone can result. In this case, the body of the instrument is excited into large-amplitude vibration as a result of being forced into vibration at its resonant frequency. The string and the body then compete for dominance and control...the body attempting to control the vibration rather than the string. Two slightly different frequencies are vying for ascendency, and energy passes back and forth from string to body. A harsh, unsteady tone with beats is the result.

There are various ways to control the wolf tone. All techniques involve shifting the wolf-note frequency from one that is commonly played to a frequency between two semitones.

THE ORCHESTRAL WOODWIND INSTRUMENTS

With the exception of their size, the orchestral stringed instruments are basically the same. This is not true for woodwind instruments. In the woodwind section of the orchestra we find open cylindrical tubes, semiclosed cylindrical tubes, and conical tubes. We also find a variety of

mechanisms for exciting an air column into oscillation: there are edge tones, single-reed tones, and double-reed tones.

The woodwinds represent a significant departure from the ideal air columns considered in Chapters 8 and 9. The finger holes in the woodwind instruments modify the resonant frequencies of the tube. By drilling a hole in a tube, we are slightly altering the shape of the tube and thereby altering the vibrational mode frequencies. The mouthpieces that are attached to the tube also influence the timbre of the instrument and do so by altering the high-frequency vibrational modes of the tube.

In our consideration of the woodwinds, we shall think of the flute, the clarinet, and the oboe as representative members. The flute is a cylindrical tube open at both ends. The piccolo is similar except it is reduced in size so that its pitch is an octave higher. The clarinet, a single-reed instrument, is a semiclosed cylindrical tube. The oboe, a double-reed instrument, is conical in shape and shares this characteristic with the bassoons as well as the saxaphones (again, a single-reed instrument). The frequency ranges of the orchestral woodwinds are given in the following figure.

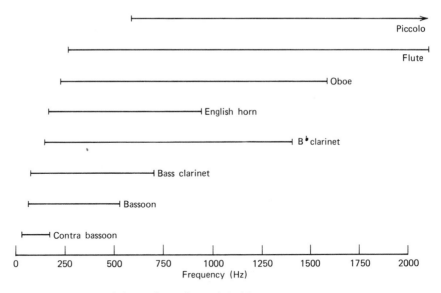

Frequency ranges of the orchestral woodwind instruments

Tubes with Side Holes The principal function of the side holes is to reduce the length of the air column so that the enclosed air vibrates at higher frequencies. If the diameter of the side hole is large enough, the side hole becomes equivalent to an open end and defines the position of a pressure node. A tiny hole—0.02 mm in diameter—would have little or no effect on the

vibrational frequencies of the enclosed air. In practice, the side holes are not equivalent to an open end because their diameters are too small. Therefore, the "effective length" of the tube is somewhat longer than the actual distance between the tube end and the side hole. The next figure shows that a tube is effectively shortened as the size of the finger hole increases.

The number of side holes required for musical purposes depends on the nature of the tube. A diatonic scale can be played on an open cylindrical tube (such as a flute) with six holes. Starting with all holes covered, the full length of the tube is utilized and the air column vibrates with its fundamental frequency. When the lowest hole is uncovered, the tube is effectively shortened and the air column vibrates with a higher frequency. As holes are successively uncovered from the lowermost hole, the frequency increases and the pitch is raised. When all holes are uncovered, the scale is one note from complete. The last note is played by once again covering all the holes and altering the blowing technique (usually blowing harder) so that the second harmonic of the tube, an octave above the fundamental, is excited. This process is referred to as "overblowing." Seven holes are required to play the scale on the clarinet. This is because the clarinet is a semiclosed cylindrical tube for which the second harmonic is missing. The third harmonic is a twelfth above the fundamental.

To continue raising the pitch beyond the first octave, the flute player continues overblowing (working from the second harmonic) and successively uncovers the lowermost hole thereby shortening the tube. The clarinet player, on the other hand, needs still more holes to carry the pitch into the second octave; in fact, 10 holes are required to carry the pitch to the *fa*, 10 whole tones above the starting tone. To play the next tone *sol*—a twelfth above the starting pitch—all the holes can once again be covered and the third harmonic excited by overblowing.

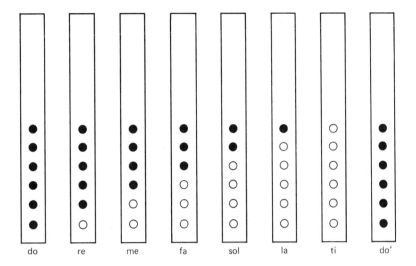

Overblowing the clarinet by lip pressure alone, however, is all but impossible. To assist the clarinetist, a speaker key is provided that covers a vent hole located approximately 15 cm from the mouthpiece. The diameter of the vent hole is too small to behave as an open end; however, its presence shifts the frequency of the fundamental and encourages the excitation of the third harmonic. The vent hole remains open for all pitches higher than sol, the twelfth above the clarinet's lowest pitch.

The oboe has a cone-shaped bore. We have learned that a conical tube is like an open cylindrical tube and has both even and odd harmonics. Therefore, like the flute, the oboe requires only six keys to play the basic scale since overblowing activates the second harmonic. However, like the clarinetist, the oboe player can overblow only with considerable difficulty; therefore assistance is provided by means of a strategically located vent hole that encourages the formation of the second harmonic.

As you might suspect, the flute player can overblow not only the second harmonic, but also the third and in some cases the fourth harmonics. In this fashion, six keys provide a range of three to four octaves. Likewise, the clarinet player can overblow the fifth harmonic and thereby extend the range of the instrument.

During our discussion an implicit assumption has been made concerning both the number of holes required to play the basic scale and the overblowing technique that employs the *same holes* to play higher pitched tones. When all the holes are covered and the full length of the tube is utilized, a pitch of frequency f_1 or overblown pitches of $2f_1$ (open cylindrical or conical tubes) and $3f_1$ (semiclosed cylindrical tube) can be played. On the other hand, when three holes are open and only part of the tube length is employed, a pitch of frequency f_1' can be played or

overblown pitches of $2f_1'$ and $3f_1'$ for the open and semiclosed cylindrical tubes, respectively. In other words, the ratio of the overblown frequency to the normally blown frequency is the same for *both* the full length and the partial length of the tube. Symbolically this can be expressed as follows:

$$\left(\frac{nf_1}{f_1}\right)_{\text{full length}} = \left(\frac{nf_1'}{f_1'}\right)_{\text{partial length}} \qquad n = 1, 2, 3, \ldots$$

This requirement is of course crucial if the same holes are to be utilized for both the normal register and the overblown register. Only certain horn shapes satisfy this requirement: they are called Bessel horns. As you might expect, both the cylindrical and conical tubes are Bessel horns and as you learned in Chapter 8, they are the only shapes that can be conveniently handled by a performer.

A cursory examination of a woodwind instrument reveals additional features of the side holes that can be understood. First, and most obvious, there are more than the six (or ten) side holes along the length of a woodwind instrument. Additional holes are present so that sharps and flats, that is, a chromatic scale, can be played. Second, the holes are not equally spaced along the length of the tube. The spacing between the holes increases as their distance from the mouthpiece increases. If a tone is being played that utilizes a length L of the tube and a new hole is covered to lower the pitch one semitone, that hole must add an increment of length approximately equal to 6 percent of L, or $0.06L$. Each time the pitch is lowered by covering an additional hole, L is increased and the next increment $0.06L$ is necessarily larger than the previous increment. Thus we can understand why the spacing between holes far from the mouthpiece is larger than it is for holes close to the mouthpiece.

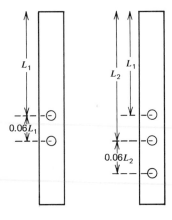

Third, for some woodwinds the hole diameters increase as their distance from the mouthpiece increases. The following figure shows a saxophone stripped of all the keys and hole covers, which makes the changing size obvious. For a tenor saxophone the bottommost hole (not visible in the figure) has a diameter of 4.45 cm while the topmost hole's diameter is 1.22 cm. Recalling that a hole is supposed to behave like an open end, it would be expected that for any tube whose cross-sectional area varies, the hole area would also have to vary. The larger the tube, the larger must be the diameter of the hole. The hole sizes in a flute—a cylindrical tube having a uniform cross section—are all essentially the same size. The holes in a clarinet, however, vary in size although the bore is cylindrical in shape. Is this a contradiction? No, it is not. Because the clarinet is closed at one end, it behaves differently than the flute. The cross-sectional area should be constant for the clarinet; however, the closed holes *add* to the effective cross-sectional area. (See the figure

(From *Horns, Strings and Harmony* by Arthur H. Benade. Copyright © 1960 by Educational Services, Inc. Used by permission of Doubleday & Co., Inc.)

below.) Since the spacing of the holes must vary to give the proper pitch variations, the hole sizes must also vary in order to keep the effective cross section uniform.

Tubes with Mouthpieces

The resonant frequencies of a tube *plus* mouthpiece is different from the resonant frequencies of the tube alone. The mouthpiece in essence adds an increment of length to the tube. For the conical tube, however, the increment of length is different for different frequencies. In particular, the mouthpiece on a conical tube "looks" longer at high frequencies than it does at low frequencies.

The mouthpiece on all reed instruments effectively closes the end of the tube to which it is attached. The natural frequencies of a semiclosed cylindrical tube were given in Chapter 8 as

$$f_n = n \frac{20.1}{4L} \sqrt{T_A} = n \frac{5.025}{L} \sqrt{T_A} \qquad n = 1, 3, 5, \ldots$$

Suppose the cavity of the mouthpiece has a volume V. Furthermore, suppose the cylindrical tube to which the mouthpiece is attached has a radius b_0. The length L of this tube having the same volume V as the mouthpiece cavity is given by

$$\Delta L = \frac{V}{\pi b_0{}^2}$$

The mouthpiece makes the tube behave as though it was increased in length by the amount ΔL. Accordingly, the resonant frequencies are given by

$$f_n = n \frac{5.025}{L + \Delta L} \sqrt{T_A} = n \frac{5.025}{L + (V/\pi b_0{}^2)} \sqrt{T_A} \qquad n = 1, 3, 5, \ldots$$

For conical bore instruments the situation is more complicated; however, an expression can be derived that closely approximates the effect of the mouthpiece. Again, let us assume the mouthpiece has a volume V and we shall further assume that V is small. If the open end of the cone has a radius b, the effective length of the cone is increased by

$$\Delta L \simeq \frac{\pi V}{b^2} n^2 \qquad n = 1, 2, 3, \ldots$$

Notice that this mouthpiece correction changes for different harmonics, that is, for different values of n. The higher the harmonic, the longer the tube behaves with the result that higher harmonics are flattened relative to the fundamental.

These considerations shed some light on double-reed instruments such as the oboe and bassoon. The space between the double reeds forms part of the mouthpiece cavity volume. However, this reed system is quite flexible and under the varying pressure of the lips the volume of the cavity can vary. Since the effective length of the instrument changes as the volume of the mouthpiece cavity changes, the flexibility of the mouthpiece accounts for the intonation difficulties of the oboe and bassoon.

The Sound Radiated from Woodwinds

In order to sustain a standing wave in a tube, energy must be constantly supplied to the air column. The performer supplies this energy input. There are two ways an air column loses energy. First, most of the energy is lost through the interaction between the air column and the walls of the tube, that is, through friction. The column of an instrument that has been played a short time is noticeably warm—evidence of friction at work. The second way in which energy is lost is doing the task for which the instrument was designed: radiating sound energy. Only about 2 percent of the energy put into the instrument is radiated away as sound. Obviously, woodwind instruments are inefficient devices.

The sound is radiated through the open holes along the tube. By far, most of the sound energy emanates from the *first* open hole below the mouthpiece. Most of the remainder of the sound energy is radiated from the second open hole. Holes beyond the second open hole can be covered and have little effect on either the amount of acoustic energy radiated or upon the pitch. The bell of the woodwind instrument is only important when all the holes are covered and the full length of the tube is being employed. For this reason, mutes have no use in the woodwind section of the orchestra.

The sound radiated from a woodwind instrument does not have the quality that would be predicted on the basis of a cylindrical or conical tube. We have already seen that both the finger holes and the mouthpiece alter the resonant frequencies of the enclosed air column. Consider, for example, the clarinet. Basically it is a semiclosed cylindrical tube and, to understand the acoustic behavior of the clarinet, the resonant properties of the ideal cylindrical tube must be understood. However, the mouthpiece cavity at the reed end, the covered finger holes, the open finger holes, and the flare in the tube at the bell end all introduce changes in the simple cylindrical symmetry of the bore.

The lowest note on a B^\flat clarinet is concert D_3 (written E_3 as the clarinet is a transposing instrument). A cylindrical tube 47.6 cm long will sound a concert D_3 when a clarinet mouthpiece is attached at one end. The resonant frequencies of such a tube have been determined and are illustrated in the figure below. The response curve shows that the resonant frequencies are odd harmonics of the fundamental frequency. The tube behaves pretty much as one would expect. The spectrum of

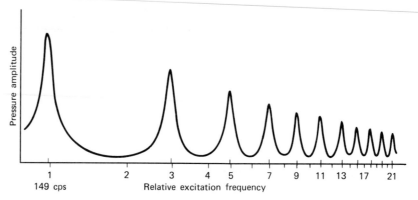

the sound radiated from such a tube is characterized by the absence of even harmonics.

The resonant frequencies of an actual clarinet can be found using the same experimental technique as that used for the 47.6 cm tube. With the results of such an experiment the sound spectrum of the clarinet can be better understood. The response curve of a clarinet is shown below where all holes are covered. The lowest resonant frequency is 153 Hz. Notice that the tube in an actual clarinet does not resonate at odd multiples of the fundamental. The fifth resonance peak, for example, has a frequency approximately equal to eight times the fundamental frequency. These are the frequencies with which the air in the clarinet bore "likes" to vibrate.

Now assume the clarinet is blown with all holes closed to produce

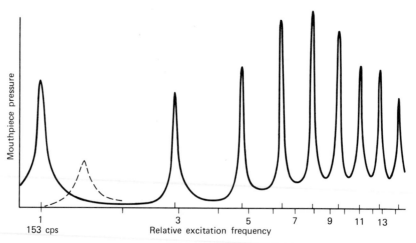

concert D_3. Since the waveform of the sound produced by a clarinet is repetitive, we can conclude that it is made up of harmonics that are exact multiples of some fundamental frequency, say 153 Hz. However, only those harmonics whose frequencies are close to the resonant frequencies of the air column will be enhanced. For this reason, the sound spectrum of the lowest note on the clarinet will show more strength in the eighth harmonic than either the seventh or ninth. The timbre of the radiated sound is certainly influenced by the departure from simple cylindrical symmetry.

Sound emanating from the side holes also influences the quality of radiated sound. In our discussion of the bell in Chapter 9 it was pointed out that high-frequency (small wavelength) sound escapes more readily than does low-frequency sound. The same can be said here: high-frequency sound is radiated more efficiently by the side holes. Thus, high-frequency harmonics will be enhanced in the perceived sound.

The resonant frequencies of other woodwind instruments can be determined in the same fashion as were those of the clarinet. The results show the same general trend, namely, the resonant frequencies are shifted somewhat from those values that would be predicted for idealized cylindrical or conical tubes.

THE ORCHESTRAL BRASS INSTRUMENTS

The bores of woodwind instruments are essentially simple shapes: either cylindrical or conical. These simple shapes, however, are complicated by the presence of side holes, a wedge-shaped mouthpiece, and a slight flare at the end. The brass instruments, on the other hand, frequently combine cylindrical sections and conical sections in the same tubular structure of the same instrument. If we think about the resonant frequencies of both a uniform cylinder and a cone, we might predict something about the resonant frequencies of the brass instruments. The complex shape of the brass instruments is not the only influencing factor on the resonant frequencies as both the mouthpiece and the bell have their effects.

The brass player has an advantage over the woodwind performer. The air column in a woodwind instrument is excited into oscillation by a vibrating reed that admits slugs of air into the instrument and thereby forces the air column into vibration. The air column, however, through feedback, strongly influences the reed with the result that the reed is virtually obligated to vibrate at the frequencies permitted to it by the air column. In the brass instruments vibrating lips excite the air column. Although feedback is important in brass acoustics, the lips are massive enough so that they can make the air column do unexpected things. A brass player can "lip up" or "lip down" a note and thereby control the air column in a manner impossible with a reed instrument.

The orchestral brass instruments consist of the trumpet, trombone, French horn, and the tuba. The trumpet, trombone, and French horn

have a lengthy cylindrical section between a rapidly flaring bell and a tapered mouthpipe. The tuba is essentially conical in form. The ranges of these instruments are shown in the following figure.

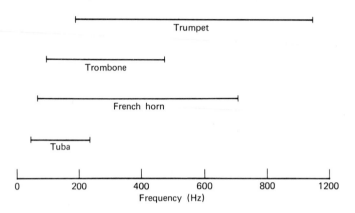

Frequency ranges of the orchestral brass instruments

TUBES, BELLS, AND MOUTHPIECES

With the exception of the tuba, the brass instruments consist of three basic sections: the mouthpiece section, which is tapered, the cylindrical part, and the bell. These sections are represented diagrammatically below. Each part actively influences the acoustical behavior of the whole.

The relative dimensions between these three parts vary from one manufacturer to another; however, the figures below are "ballpark" figures and provide a useful frame of reference. The data below refer to the instruments minimum length (slide in or valves open).

	Trumpet	Trombone	French Horn
Mouthpiece section	21%	9%	11%
Cylindrical section	29%	52%	61%
Bell section	50%	39%	28%
Total length	140 cm	264 cm	528 cm

The physics of the brass instruments poses interesting problems that are by no means completely understood. We shall consider the trumpet in some detail and in so doing we shall recognize that the brass instruments are complicated acoustical systems. A trumpet is a tube closed at the lip end. The air column is most easily excited at one of its resonant frequencies and through feedback "tells" the vibrating lips what to do. Therefore, the blown frequencies coincide very closely to the resonant frequencies of the air column. The approximate blown frequencies of the B^{\flat} trumpet—with all valves open—are given in Table 1. In the second column of the table it is seen that the blown frequencies are approximately multiples of 115 Hz. However, the fundamental in the series, namely 115 Hz, is missing.

Table 1 Trumpet Frequencies

Blown Frequencies (Hz)	Breakdown of Frequencies
231	$231 \approx 2 \times 115$
346	$346 \approx 3 \times 115$
455	$455 \approx 4 \times 115$
570	$570 \approx 5 \times 115$
685	$685 \approx 6 \times 115$

Already there are problems. We have come to expect that a tube closed at one end will have only odd harmonics. Remembering, however, that the bore of a trumpet combines both cylindrical and conical sections, this result is somewhat more tenable. The missing fundamental raises another question that may point to a problem.

The overall length of a trumpet is in the 140 cm range. Suppose we start with a cylindrical tube 140 cm long and approximately 0.6 cm in radius. The resonant frequencies of the air column in such a trumpetlike tube can be determined. The results follow closely the expectations of Chapter 8. If the tube is closed at one end, the resonant frequencies are odd multiples of about 62 Hz, which leads to harmonic frequencies of 62, 186, 310, 434, 558, 682 Hz, etc. These frequencies are close, but not close enough, to the playing frequencies of a trumpet. (See figure (a) on page 122.)

If a trumpet bell section is added to this trumpetlike tube, the frequencies of the resonances are altered. The lowest resonant frequency is not changed significantly, but higher resonant frequencies are shifted progressively downwards with the result that the difference between successive frequencies is reduced. (See figure (b) on page 122.) The reason for this is that the standing wave corresponding to the fundamental is reflected from essentially the same "end" as it was before the bell section was added. Standing waves of higher frequencies,

however, penetrate deeper and deeper into the bell section before they are reflected with the result that the effective length of the tube plus bell section is longer the higher the frequency.

The mouthpiece section of a trumpet consists of a small cuplike volume connected to a tapered tube. If the mouthpiece section is added to our 140 cm trumpet-like tube (without the bell section), all resonant frequencies are lowered. In fact, the influence of the mouthpiece on the resonant frequencies of the brass instruments is analogous to the corresponding influence of the mouthpiece in the woodwinds. When the wavelength is long compared to the length of the mouthpiece section, which is the prevailing situation at low frequencies, the tube is lengthened by an amount equal to a length of tube having the same volume as the added mouthpiece section. For example, suppose the volume of the mouthpiece section is $7 \, \text{cm}^3$. A length of 6.2 cm of our trumpetlike tube (radius 0.60 cm) has a volume of $7 \, \text{cm}^3$. Therefore, at low frequencies the mouthpiece section adds 6.2 cm to the 140 cm length. The resonant frequencies are correspondingly lowered.

At higher frequencies (wavelength less than the mouthpiece dimension), the mouthpiece section, cuplike volume plus tapered section, has its own resonance characteristics. These characteristics make the tube look longer to high frequencies than it does to low frequencies. In fact, the higher the frequency, the longer the tube-plus-mouthpiece-section appears.

We have seen that each component of the brass instrument exerts its own influence on the resonant frequencies of the enclosed air column. If the length of the cylindrical section is chosen properly, obviously less than 140 cm, the combined influence of the bell section and mouthpiece section can bring the resonant frequencies to acceptable values. A resonance curve is shown for a trumpet in the figure below. From this

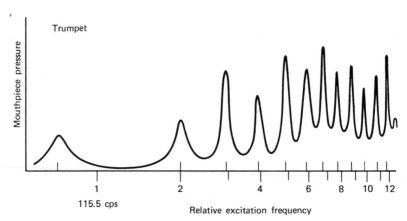

(Reprinted from *The Acoustical Foundations of Music* by John Backus. By permission of W. W. Norton & Co., Inc. Copyright © 1969 by W. W. Norton & Co., Inc.)

figure it is seen that f_2, f_3, f_4, etc. are harmonics of a fundamental that *doesn't exist*. The lowest actual resonance is almost a fifth below the *fictitious fundamental*. This lowest frequency resonance is referred to as the *pedal tone*. The pedal tone is out of tune with the other harmonics; however, it doesn't matter since that resonance is never used. Finally, higher harmonics are slightly flat; the performer must compensate for this by pulling the tone up slightly.

We now have an instrument that can play on several harmonic frequencies. Now we shall fill in the tones between these resonances so that a chromatic scale can be played.

SLIDES AND VALVES

The harmonic frequencies of a brass instrument are determined by the length of the air column. If the length is changed, the frequencies are changed. The most straightforward way of altering the length is accomplished by the trombone. Other brass instruments use a system of valves.

Whether a slide or a valve system is employed the same task must be accomplished. Namely, the interval between the harmonics f_2 and f_3 (the first two resonances used by a brass instrument) must be bridged. The interval f_2 and f_3 is a musical fifth with six semitones between. Thus there must be six definite lengths each giving rise to successive semitones between f_2 and f_3. Let's see how this works.

We shall start with the trombone. The trombone is equipped with a slide that can continuously increase the length of the instrument from its minimum to its maximum length. Between these extremes there are five positions that the trombone student quickly learns. These six positions, the five intermediate plus the maximum length, define six lengths that bridge the gap between f_2 and f_3. When the slide is all the way in (the first position), the trombone has resonant frequencies of *approximately* 116, 174, 233, 292 Hz, etc.

To lower the frequency by one semitone is to decrease the frequency by about 6 percent. Therefore, the effective tube length must be increased by about 6 percent. With the slide in the first position, the overall length of the instrument is approximately 270 cm; therefore, to lower the pitch one semitone, a length of $0.06(270 \text{ cm}) \simeq 16 \text{ cm}$ must be added to the instrument. This can be accomplished by moving the slide *out* 8 cm to the second position. In the second position a new family of resonances can be utilized. Their frequencies are approximately 110, 165, 220, 277 Hz, etc.

In the second position the instrument length is about 286 cm. To lower the pitch another semitone, a length of 17 cm must be added to the instrument ($0.06 \times 286 \text{ cm}$). Moving the slide out 8.5 cm to the third position does the trick. In this position a new set of resonant frequencies can be excited. In similar fashion, each succeeding position of the slide

adds a definite length to the instrument, lowers the pitch by a semitone, and makes available a new set of resonances. The figure below illustrates the total process. As it can be seen, the gap between f_2 and f_3 has been closed.

(Reprinted from *The Trumpet and Trombone* by Philip Bate. By permission of W. W. Norton & Co., Inc. Copyright © 1966 by Philip Bate.)

The use of valves to close the f_2 to f_3 interval is more complicated. For example, let us consider the trumpet. Like some other brass instruments, the trumpet has three valves. We shall assume the trumpet's overall length is 140 cm when all valves are open. When the second valve is depressed, the length is increased by 6 percent, or by 8.4 cm, and the pitch is lowered by one semitone. When the first valve is depressed a new tube length is added to the trumpet, which lowers the pitch two semitones. What additional length is required? When valve one is depressed a length of 8.4 cm + 0.06(148.4 cm) or 17.3 cm must be added. So far so good as Table 2 illustrates.

The next step is not so clear-cut. In order to lower the pitch three semitones both valves one and two are depressed. Valve one adds 17.3

Table 2 Valve Action of the Trumpet

Note Sounded	Valve	Semitone change	L(ideal) (cm)	L(actual) (cm)
B♭	0	0	140	140
A	2	1	140 + 8.4 (148.4)	148.4
A♭	1	2	148.4 + 8.9 (157.3)	157.3
G	1,2	3	166.7	165.7

and valve two adds 8.4 cm for a total added length of 25.7 cm or an overall length of 165.7 cm. What *should* the overall length be? It should be $157.3 \text{ cm} + 0.06(157.3 \text{ cm}) = 166.7 \text{ cm}$. Here is a discrepancy. When valves one and two are depressed the actual overall length is 165.7 cm while it should be 166.7 cm. As a result the sounded note is somewhat sharp. There are two responses to this situation. The performer can lip the tone down. Or the performer can activate a trigger mechanism that slides a small tube section out and thereby adds to the overall length of the instrument.

With combinations of the three valves, the pitch can be changed a semitone at a time. With each change a new family of harmonics becomes available and the f_2 to f_3 interval is bridged; however, as we have observed, the three valves cannot do the job perfectly and so compromises must be made.

INSTRUMENTS OF THE BRASS SECTION

We have already discussed the trumpet and trombone. The resonance curve of the trumpet has already been shown and the corresponding curves for the trombone and French horn are shown below. As with the trumpet, the first resonance of the trombone is out of tune with the other resonances; however, once again this resonance is not employed. For both the trumpet and the trombone only eight to nine resonances are employed. This is not so with the French horn, which utilizes resonances up to the sixteenth. This explains the very slippery intonation of the French horn. As can be seen from the curve, the resonant frequencies move very close together at high frequencies. If a French horn player wants to excite the fourteenth resonance frequency, the embouchure must be exact and the lip tension precise. Otherwise, the thirteenth or the fifteenth may be mistakenly and temporarily excited. Novice (and professional) players have difficulties "hitting the right note."

Another significant difference between the French horn and the other brass instruments is the mouthpiece. The trumpet and trombone mouthpieces have cuplike structures with an edge at the entrance to the tapered section. The mouthpiece of the French horn, on the other hand, has neither a cup shape nor an edge but has a gradual and smooth transition from the lip rim to the back bore. These differences in the mouthpieces affect the timbre of each instrument; however, the acoustical basis for understanding this has yet to be worked out.

The tuba is the low-voiced member of the brass family. It is a valve instrument. However, because of the great length of the tuba, the problems inherent with the valve system are magnified with the tuba. As a result, some models of the tuba have four valves, others five valves, which are included to keep the discrepancies to a minimum.

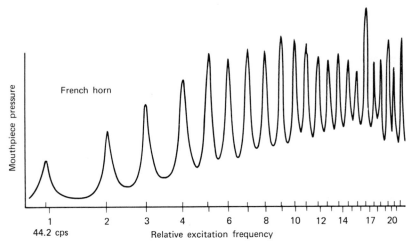

(Reprinted from *The Acoustical Foundations of Music* by John Backus. By permission of W. W. Norton & Co., Inc. Copyright © 1969 by W. W. Norton & Co., Inc.)

Trumpet

French horn

The acoustic energy is radiated entirely from the bell of the brass instrument. This means that the sound emanating from a brass instrument is more directive, particularly at the high frequencies. This also means that mutes can be used with these instruments. Mutes are hollow devices that have resonance properties of their own. Those frequencies radiating from the bell that are near the resonant frequency of the mute are strongly absorbed and significantly alters the timbre of the resulting sound.

THE PIANO

The piano took its present form in 1855 when Henry Steinway, an American piano manufacturer, introduced the cast-iron frame, which gave to this popular instrument a brilliance and power unobtainable by its predecessors. The clavichord, one of the ancestors of the piano, was a fifteenth century accomplishment. Like the piano, the clavichord is a string-percussion instrument in which a metallic wedge strikes a taut string and initiates a vibration. The sound of the clavichord is anemic.

The harpsichord is a more recent forerunner of the piano. The strings of this instrument are set into oscillation by a quill that plucks the strings. More than one string for each note was introduced to increase the volume. A shortcoming of the harpsichord is that the plucking mechanism does not permit the execution of dynamics; that is, it cannot be played softly or loudly. Also, the intensity of sound is weak.

The piano overcomes these shortcomings.

The Structure of the Modern Piano

The piano embraces over seven octaves—88 semitones, 88 keys—starting with A_0, 27.5 Hz and ending with C_8, 4186 Hz. The bass strings derive their low frequency as a result of their long length and their great massiveness, the latter achieved by overwinding one wire on another. The 10 lowest pitched notes of the piano have one string each. The next 18 notes have two strings each and the remaining 60 notes have three strings each. The total tension of the tuned strings can be as large as 60,000 pounds in the case of a concert-grand piano. It is the cast-iron frame that enables the instrument to stand up under these forces.

The strings are stretched between a tuning peg at the keyboard end, over a bridge, and anchored to the cast-iron frame at the other end. To stay in tune the strings must maintain their tension. In other words, the tuning pegs must not yield to the forces exerted upon them by the strings. To give them the stability they require, the tuning pegs are set into a pin block made up of as many as 41 layers of cross-grained hardwood. The bridge is mounted on the soundboard, generally made of Norway spruce.

The piano string is set into vibration by the percussive impact of a hammer. A little reflection on this matter will reveal that when a piano key

(By permission of Steinway and Sons.)

is struck, a very elaborate chain of events is initiated. Think of what has to happen if a musical score calls for striking the same key several times in rapid succession. First, the hammer must rebound from the string immediately after impact so that it does not damp out the ensuing vibrations of the string. After striking the string, the hammer must be held so as not to bounce up and hit the string a second time. Then the striking mechanism must reset itself quickly so that the next note in the rapid succession can be played. When a musical score calls for different keys played staccato in rapid succession, the striking mechanism must damp out the vibrations of one set of strings before the next key is struck. In other words, the sound must be initiated when the key is depressed and terminated when the key is lifted.

CAST-IRON FRAME

PIN BLOCK

SOUNDBOARD

CASE

BRIDGE

PEDALS

ACTION

KEYBOARD

The mechanism that meets these demands is called the *action* of the piano. It is an intricate and complicated mechanical system. Some pianos have over 7000 separate parts in the action, which means approximately 80 parts per key.

The figure on page 164 shows the action in a modern piano. The hammer is made of felt and can neither be too soft (leading to a lackluster tone) nor too hard (leading to a harsh tone). When the key is

1	Keybed	14	Dag	27	Support Flange	40	Hammer Rest	53	Underlever
2	Keyframe	15	Key	28	Support	41	Regulating Rail	54	Underlever Top Flange
3	Front Rail	16	Key Covering	29	Support Cushion	42	Letoff Regulating Screw	55	Damper Wire Screw
4	Front Rail Pin	17	Sharp	30	Fly	43	Hammer Rail	56	Tab
5	Front Rail Pin Punching	18	Front Pin Bushing	31	Tender	44	Hammershank Flange	57	Damper Stop Rail
6	Balance Rail	19	Key Button	32	Fly Regulating Screw	45	Drop Screw	58	Damper Wire
7	Balance Rail Stud	20	Balance Pin Bushing	33	Spoon	46	Hammershank	59	Damper Guide Rail
8	Balance Rail Pin	21	Capstan Screw	34	Support Top Flange	47	Knuckle	60	Damper Head
9	Balance Rail Bearing	22	Backcheck	35	Balancier	48	Hammer	61	Damper Felts
10	Back Rail	23	Backcheck Wire	36	Balancier Regulating Screw	49	Underlever Frame	62	String
11	Back Rail Cloth	24	Underlever Key Cushion	37	Repetition Spring	50	Underlever Frame Spring	63	Agraffe
12	Key Stop Rail Prop	25	Action Hanger	38	Repetition Felt	51	Underlever Frame Cushion	64	Tuning Pin
13	Key Stop Rail	26	Support Rail	39	Balancier Covering	52	Underlever Flange	65	Sostenuto Rod

(By permission of Steinway and Sons.)

164

depressed (see the illustration), the hammer is set into motion and the damper is lifted from the strings. Just before the hammer strikes the string, the jack regulator essentially decouples the hammer from the key and the hammer, already moving rapidly, continues freely toward the string. (**Note.** When a key is depressed very slowly, the hammer does not achieve the momentum necessary to carry it to the string and no sound is heard.) The rebounding hammer is caught and held by the backcheck. When the key is lifted, the damper comes back to rest on the string.

There are three pedals on the piano. The pedal seen on the right when facing the instrument is the damper pedal. When this pedal is depressed, all dampers are disengaged so that the strings are free to vibrate. Releasing the pedal returns the dampers to the strings. The left-hand pedal is the soft pedal. When it is depressed, the entire action is shifted so that the hammer, instead of striking all the strings of a note, strikes only one or two strings thereby decreasing the sound intensity. The middle pedal sustains only the tones that are played at the time the pedal is depressed.

The Sound Producing Mechanism of the Piano

The origin of sound in the piano is the vibrating string. We discovered in Chapter 8 that the frequencies with which a string vibrates is given by

$$f_n = \frac{n}{2L} \sqrt{\frac{\mathcal{T}}{d}} \qquad n = 1, 2, 3, \ldots$$

The greater the tension \mathcal{T} of the string, the faster the string snaps back from its displaced position to its equilibrium position, and the greater are the frequencies of its normal vibrational modes. The above relationship is correct for strings having no stiffness. Piano strings have inherent stiffness that contributes to the swiftness with which a string will snap back toward its equilibrium position.

In other words, a string with inherent stiffness will vibrate with a higher frequency than is predicted by the above relationship. Moreover, the effect of the stiffness is more pronounced on the higher harmonics than it is on the fundamental or low harmonics. In fact, the increase in frequency is related to the square of the frequency so that $2f$ (the octave) will be shifted four times more than the fundamental frequency f. These considerations suggest that we should speak of overtones for a vibrating piano string rather than harmonics since frequencies of the higher vibrational modes are not whole-number multiples of a fundamental frequency.

Two strings having the same densities and tensions but differing in their lengths by a factor of two would differ in their frequencies by a factor of two. Their pitches would differ by an octave. If the lengths of the piano strings were halved for each octave rise in pitch, the strings of

the highest pitched notes would be too short. Therefore, the length ratio for the octave is not 2:1, but ranges from 1.88:1 to 1.94:1. This ratio is called the *stringing scale* of the piano. The departure of the stringing scale from the 2:1 ratio is compensated by both varying tensions and densities so that the frequency ratio of the octave is indeed 2:1!

As with the violin, a vibrating piano string transfers very little acoustic energy to the surrounding air. The vibrations of the string are therefore transferred via the bridge to the soundboard and then radiated as acoustic energy to the air. The piano string forces the soundboard into oscillation.

We concluded Chapter 2 with the question concerning the physical requirements of a soundboard. Do you know the answer? Clearly, the ideal soundboard would have a uniform response across the frequency range of the piano. A soundboard with a strong response for C_4, a weak response for A_4, etc. is unacceptable.

The Character of the Sound Transmitted by the Piano

In Chapter 13 we shall learn that there are different ways of building a musical scale. The scale used for the modern piano is the *tempered scale*. In this scale the octave is divided into 12 equal frequency intervals.

Let us consider some of the factors that contribute to the complex sound radiated from the piano.

(i) Inharmonicities

The stiffness of piano strings destroys the harmonic character of its sound. This inharmonicity makes the piano's sound distinctive. Juries of both musicians and nonmusicians have been asked to identify pianolike tones produced synthetically. With overwhelming consistency only those tones were selected that correctly introduced inharmonicities to the sound.

Piano sounds have been carefully analyzed. As we would predict from our earlier discussion, the higher the frequency of the overtone for any note on the piano, the more it departs from its harmonic counterpart. The overtones become sharper and sharper as their frequency increases. The overtone character of the lowest note of the piano, A_0, is shown in the graph on page 167. If the complex tone was made up of harmonics, it would follow the straight line. The curved line identifies the frequencies of the overtones. Obviously the tone is inharmonic.

(ii) Attack and Decay

As we have already seen the attack and decay of a tone are characteristic of an instrument. When a hammer strikes a piano string, a tone is initiated in percussive fashion. The resulting sound builds to a maximum and then dies away. The decay of a piano tone is a characteristic

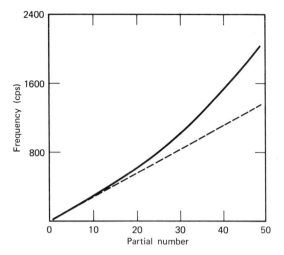

property of its sound. The graphs on pages 168 and 169 show the decay of C_1. These curves show that the overtones of C_1 all have their own decay rates. Some overtones like the fifth decay and then build up again to increased loudnesses. Others, like the second and fifteenth overtones decay more uniformly. Thus the timbre of a piano tone changes as it decays.

(iii) Beats

"Warmth" of a musical tone is a subjective quality. A tone played by several violins in unison is judged "warm" while the same tone played by one violin is described as lacking "warmth." It has been suggested that "warmth" originates with beats. Several violins would rarely be exactly in tune and a slow beat would result.

The idea of beats receives further support by the fact that synthetic tones built up from perfectly tuned harmonics lack warmth. For many notes of the piano the hammer strikes more than one string. If these strings are even slightly out of tune, a slow beat results. This beat phenomenon is believed to provide the warmth attributed to the piano tone. If the beats become too prominent, they are annoying and the services of a piano tuner are needed.

(iv) Mechanical noise

When a note on the piano is played, some sound is produced that does *not* originate with the vibrating string. The impact noise of the hammer striking the string is particularly noticeable in the upper register of the instrument. Pedal noise contributes to the overall acoustic effect. Unstruck strings can vibrate sympathetically with a struck string. All of these "incidentals" are an important part of that sound we identify with the piano.

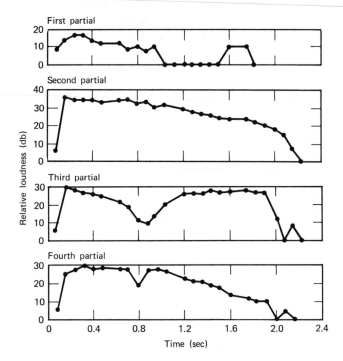

(Reprinted by permission from the Journal of the Acoustical Society of America.)

REFERENCES

1. Hutchins, Carleen M. and Fielding, Francis L., "Acoustical Measurement of Violins," *Physics Today* (July 1968), p. 35.

This paper describes experiments to determine the resonance properties of the violin.

2. Hutchins, Carleen., "Founding a Family of Fiddles," *Physics Today* (February 1967), p. 23.

3. Hutchins, Carleen M., "The Physics of Violins," *Scientific American* (November 1962), p. 78.

4. Backus, J., "Vibrations of the Reed and the Air Column in the Clarinet," *JASA* **33**, 806 (1961).

5. Bate, Philip, "The Trumpet and Trombone," Norton, New York, 1966.

6. Benade, A. H., "The Physics of Brasses," *Scientific American* (July 1973).

7. Backus, John, "Input Impedance Curves for the Brass Instruments," *JASA* **60**, 470 (1976).

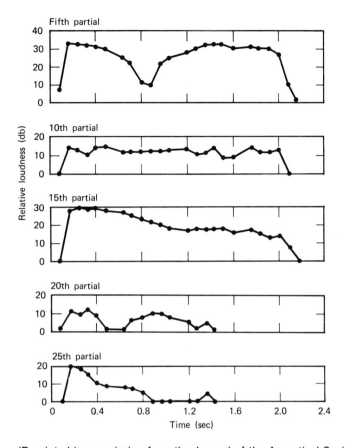

8. Blackham, E. D., "The Physics of the Piano," *Scientific American*, (December 1965).

For Further Thought

1. The viola is tuned a perfect fifth below the violin. From your knowledge of string vibrations, how can this lowering of pitch be achieved? If you desired to preserve as much of the "violin character" as possible, how would you proceed in lowering the pitch of the viola?

2. The open D-string on a violin is tuned to a frequency of 294 Hz.
 (a) What would be the desirable frequency of the MAR?
 (b) Let's use this frequency to estimate the volume of the enclosed air in a good violin. To carry out this estimation, assume that the thickness of the top plate at the f-holes is 0.25 cm and that the

area of each f-hole is 6.5 cm². What is the calculated volume? Does your result seem reasonable?

3. The frequency of the air resonance is given by the formula

$$f = \frac{v}{2\pi} \sqrt{\frac{A}{lV}}$$

If all the dimensions of the cello were exactly twice those of a violin, show that the air resonance of a cello would be one-half that of a violin.

4. Explain why the flute player can accomplish with six side holes what a clarinetist requires seven side holes to achieve; namely, play the sequence of notes making up the diatonic scale.

5. How many side holes must a flutelike instrument have if *two* octaves are to be played? How many side holes must a clarinetlike instrument have if *two* octaves are to be played?

6. The mouthpiece of a clarinet forms a closed end. Explain how it is that the air column exercises a controlling influence on the vibrations of the reed.

7. Assume the B♭ clarinet has an overall length of 67 cm.
 (a) Based on this actual length, what should be the lowest frequency of the clarinet?
 (b) The lowest note played by the B♭ clarinet is D_3. What is the effective length of the clarinet?
 (c) How do you account for the answers in (a) and (b)?

8. Why are there more keys on the clarinet than there are on the flute?

9. Let us assume that the length of the trombone is 270 cm when the slide is in the first position. How far would the slide have to be moved out to lower the pitch by four semitones?

10. Assume that the mouthpiece of a clarinet has a volume of 8.0 cm³. The mouthpiece slides on to a bore having an inner radius of 0.75 cm. Discuss how the addition of this mouthpiece would alter the resonant frequencies of the tube to which it is attached.

11. The third valve of a trumpet adds a length of tube to lower the pitch by three semitones. (Valve three alone is more-or-less equivalent to valves 1 and 2 together.) Consider the situation when all three valves of the trumpet are depressed. What discrepancy would you predict between the actual and ideal lengths?

12. Consider a mute with the dimensions shown below. Determine how this mute influences the timbre of the transmitted sound of the

trumpet. The volume of the truncated cone is given by $V = \frac{1}{3}\pi h\,(a^2 + ab + b^2)$.

$h = 15\ \text{cm}$
$a = \ \ 5\ \text{cm}$
$b = \ \ 1\ \text{cm}$
$l = \ \ 1\ \text{cm}$

13. A piano string is 140 cm long with a linear density of 0.042 kg/m. The string is stretched to a tension of 2000 newtons. The hammer is designed to strike the string a distance equal to $\frac{1}{7}$ of its length from one end. What is the maximum time that the hammer can be in contact with the string if we assume that the hammer must break contact with the string before the wave reflected from the far end of the string returns to the point of hammer impact.

12 The Musical Environment

Ideally, music is a form of dialogue: an exchange of ideas between performer, listener, and the environment. If effective communication is to be achieved, the performer, the listener, and the environment must actively participate in the dialogical process. The alternative is an ineffectual monologue that leaves all participants dissatisfied.

THE INFLUENCE OF THE ENVIRONMENT

On the Performer

A prerequisite for a balanced orchestra is the ability of the performers to hear each other. Therefore, sounds must be carried from one side of a stage to the other. The surfaces of the stage, their composition and shape, (sides and top) either enhance the ability of performers to hear each other or detract from it. The term *ensemble* refers to performers' ability to play together in unison. Good ensemble requires good acoustics.

A concerto is a large-scale work in several movements for a solo instrument and orchestra. The focus of attention in this situation is on the soloist. It is desirable to achieve a good balance between the soloist and the orchestra. The conductor is responsible for controlling the orchestra so that the soloist can be heard; however, once again the acoustical setting can be either an asset or a hinderance. Isaac Stern has pointed out that in the proper acoustical environment a "friendly aura" surrounds each note with each note blending into the previous sound thereby giving him sound to work with. Mr. Stern has written to me that Powell Symphony Hall in St. Louis "provides that alive and vibrant quality which makes a fine hall an additional instrument which the artist can use in his performance."

173

Conductors who work regularly in a particular concert hall become intimately familiar with its acoustical idiosyncracies. In a concert hall where sounds tend to die quickly, the conductor may compensate by developing techniques in which the attack and release of each tone are prolonged. Conversely, in a concert hall that imparts fullness of tone to each note, the conductor may develop a style that approaches the attack and release of each note more abruptly. This is one reason why a conductor visiting a strange concert hall may not live up to his reputation.

On the Audience

The concertgoer wants the experience of being surrounded by uniform sound possessing a balance between the low and high frequencies, a blending of one tone with another, and a marked distinction between the pianissimo and fortissimo passages—regardless of where he or she sits! Once again, the environment must play an active part in this dialogue if the listener's goal is to be achieved. Concert halls must be designed to minimize audience noise while at the same time supporting the sound that emanates from the orchestra. Echos must surely be avoided. Sound that is reflected from various surfaces of the auditorium must reach the ear of the listener soon after the direct sound. Otherwise the direct sound and reflected sound can become dissociated with chaotic confusion as the result.

Listeners may also derive more satisfaction from one style of music played in a particular concert hall than from a second style of music. As we shall see, musical composition and architecture have influenced each other. Composers in a particular period have had certain structures available to them and in many instances they wrote with a specific acoustical environment in mind.

On the Critic

A critic can be thought of as a very refined listener. When a new concert hall is built or an existing one modified, the architects and acoustical consultants await the reaction of the critics with great anticipation—as well as with anxiety and perhaps foreboding. The critic may write, as Frank Howes of The Times, London once wrote: "This Hall is probably the worst in the World" Or, the critic may respond, as Cyrus Durgin of the Boston Globe did: ". . . the new dark-gray, acoustic stage shell has greatly improved the sound The difference is extraordinary."

In order to set the stage for our acoustical consideration in this chapter, let us reflect on the commentary of some critics taking special note of their language.

> A *full, well-blended,* yet *detailed* sound is projected to the furthermost benches . . . orchestra tuttis can be heard with surprising *clarity* . . .
> Eric McLean, Montreal Star

The sound in Philadelphia is beautifully *blended*. There is not as much *brilliance* there as in Boston; the sound is sort of velvety. I enjoyed this hall right away; it is more *intimate* than Carnegie.

Cyrus Durgin, Boston Globe

Compared with Carnegie and Boston, it is the most mellow, the *warmest*. It is not as *reverberant* . . . has a glorious *brilliant* tone.

Howard Taubman, New York Times

The tone, though well *balanced*, tended to be dry and thin. It was a bit richer upstairs. On the whole, *mellowness* and *warmth* were absent.

Howard Taubman, New York Times

Two kinds of music seem to suffer the most; the elusive atmospheric piece, and the sumptuous late Romantic scores of Wagner, Strauss and Rachmaninoff

D. Shawe-Taylor, Sunday Times,
London

The critic employs an idiom that faithfully conveys their response to an acoustical environment. I have italicized words that are commonly used in their critiques. Although these words, and others like them, have a subjective quality about them, they can be correlated with objective, quantitative acoustical factors. These acoustical factors are discussed later in this chapter.

ACOUSTICS— ART OR SCIENCE

With the uncertainties that plague architects and acousticians prior to the opening of a concert hall on the one hand, along with the many myths that have arisen to account for the acoustical properties of existing concert halls on the other hand, one is tempted to ask the question: is acoustics a science or is it an art? The question is further compounded because acoustics is both old and new. Viewed in one way, acoustics is one of the oldest branches of physics. Over 2500 years ago the mystical Pythagoras showed that a vibrating string gave rise to harmonious intervals when the string lengths were related by simple whole numbers. This discovery convinced Pythagoras that numbers provided the clues to the mysteries of nature. Later, acoustics played a role in the design and location of the open-air theaters used by the Greeks and Romans.

These examples, however, are somewhat misleading. Up until 1900, the design of an acoustically successful concert hall was a matter of good fortune. It has only been since 1900 that acoustical knowledge has been applied to the design of music halls and auditoriums. Even now, the acoustician often plays second fiddle to the architect; that is, aesthetic and utilitarian concerns often supersede the considerations of the acoustical design consultant.

These considerations enable us to understand some of the myths that are a part of the acoustical folklore. It is commonly believed that the

acoustical qualities of a concert hall improve with age. This is a myth whose half-truth character has helped perpetuate it. Yes, old concert halls are likely to have very fine acoustics; however, they are old *because* they have good acoustics and not vice versa. Communities grow weary of concert halls with poor acoustics and these structures become the victims of the headache ball. Witness, for example, the nineteenth century Italian Opera House in New York City. Its acoustics were so poor it was destroyed six years after it was built. In other words, old halls with good acoustics had their good acoustics from their youth.

In an attempt to "explain" the fine acoustics of concert halls, attention has often been focused on an unusual characteristic of either the structure itself or the immediate environs. A dry well or broken wine bottles under the stage have been given credit for a concert hall's fine acoustics. Wooden surfaces or even gold paint on statuary have been cited as the secret of a hall's acoustical excellence. Modern construction materials such as concrete, glass, and steel have been blamed for poor acoustics. These kinds of considerations camouflage rather than clarify the pertinent issues.

Is architectural acoustics a science or an art? Perhaps it is most appropriate to respond that acoustics is a developing science with questions that remain to be answered and understood. An unequivocal testimony to the skill and understanding of the acoustician is found in the ability to transform an acoustically inferior hall into an adequate or very good one. (Avery Fisher Hall in the Lincoln Center was transformed by acoustician Cyril M. Harris and architect Philip Johnson. On October 19, 1976, the opening of the new hall featured movements from Mahler's Symphony No. 9. The critics raved!) Furthermore, if acoustical considerations are made a determining part of the initial planning of a structure, the outcome can be anticipated with confidence.

BEHAVIOR OF SOUND— A REVIEW AND AN EXTENSION

In Chapter 3 it was established that the behavioral characteristics of sound waves include reflection, diffraction, and interference. Let us reconsider these behaviors and refine our understanding.

Reflection ($\lambda \ll x$)

A sound wave incident upon a hard plane surface is reflected by that surface such that the angle of reflection equals the angle of incidence. This statement can now be refined; namely, if the surface dimension, x, is large compared to the wavelength, λ, of the incident sound, regular reflection occurs. Specifically, for wavelengths such that $x \geq 4\lambda$, regular reflection is observed.

Surfaces can be used to control and distribute sound energy. A convex-shaped reflecting surface distributes the sound over a large area.

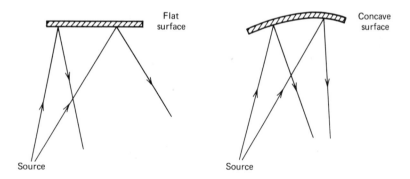

Convex
shape

Source

A flat reflector does not distribute the sound nearly as much. A concave-shaped reflecting surface focuses the sound. Such concave surfaces can be troublesome and should be avoided in the design of enclosures.

Flat
surface

Source

Concave
surface

Source

Diffusion (λ ≃ x)

When $\lambda \ll x$, regular reflection results. When dimensions on a surface approximately equal the wavelength, that is, when $x \simeq \lambda$, the sound is scattered, by reflection, in all directions. This scattering is called *diffuse reflection.* (See question 1, 193.)

Diffraction (λ ⩾ x)

When elements of a surface have dimensions such that $\lambda \geqslant x$, diffraction becomes the important behavior. We shall see later how this becomes important in acoustical considerations.

Absorption

Sound energy is absorbed both by the air through which it propagates and by the surfaces against which it comes into contact. High frequencies are absorbed by air more readily than are low frequencies. It is for this reason that nearby thunder cracks, while distant thunder, with all the high frequencies absorbed by the intervening air, rumbles. Air absorption is not too important in architectural acoustics except in the very largest auditoriums.

Surface absorption is another matter. By a mere change in the nature of the surface material the acoustical properties of an environment can be completely altered. In general, hard, smooth surfaces reflect most of the incident sound while soft, porous surfaces such as fabrics absorb a large percentage of the incident sound energy. The absorption properties of a surface can be expressed quantitatively.

The larger the area of a surface, the more sound energy it will absorb. In fact, the amount of energy absorbed is directly proportional to the surface area. We can therefore write

$$a \propto s$$

or

$$a = \alpha s$$

where

a = absorptivity

s = the area of the surface in square meters (m^2)

and

α = the absorption coefficient

The absorption coefficient α is a number representing the fraction of sound energy absorbed by a surface. Its value can range from 0 (no sound energy absorbed) to 1.0 (all incident sound energy absorbed). Since surfaces absorb differently at different frequencies, the absorption coefficient must be determined for several different frequencies.

Let us consider three surfaces each having dimensions of 1.25 × 1.60 meters. The area of each surface is 2.0 m^2. Furthermore, let us assume sound having a frequency of 500 Hz is incident upon each surface. No sound is reflected through an open window; therefore, $\alpha = 1.0$ and $a = 2.0$ m^2. An open window is a perfect absorber. Let the second surface be plaster over 10 cm of brick. For this surface, $\alpha = 0.02$;

that is, 2 percent of the sound energy is absorbed. Therefore, the absorptivity of this surface is given by $a = 0.02(2 \text{ m}^2) = 0.04 \text{ m}^2$. Finally, consider a surface covered with 3.8 cm of thick fuzz. In this case $\alpha = 0.78$ meaning that 78 percent of the incident sound energy is absorbed. The absorptivity is $a = 0.78(2 \text{ m}^2) = 1.56 \text{ m}^2$.

The total absorption A of an enclosure can be determined by treating all exposed surfaces in the above manner and summing the individual absorptivities; that is,

$$A = \alpha_1 s_1 + \alpha_2 s_2 + \alpha_3 s_3 + \cdots$$

In the above expression, α_1 is the absorption coefficient of a surface having an area s_1. Absorptivities are expressed in *sabins* rather than square meters; that is, the fuzz surface above has an absorptivity of 1.56 sabins.

1.25 m

Incident

sound

1.60

Open
window

$\alpha = 1.0$ $a = 2.0 \text{ m}^2$
 $= 2.0 \text{ sabines}$

Incident

Reflected

Plaster on
10 cm
brick

$\alpha = 0.02$ $a = 0.04 \text{ m}^2$
 $= 0.04 \text{ sabine}$

Incident

Reflected

3.8 cm
Thick fuzz

$\alpha = 0.78$ $a = 1.56 \text{ m}^2$
 $= 1.56 \text{ sabines}$

Let us now determine the total absorption for a hypothetical enclosure and illustrate at the same time the effect of surface materials. Let us assume a room with dimensions of $4\,\text{m} \times 4\,\text{m} \times 4\,\text{m}$ having concrete surfaces. The absorption coefficient for poured, unpainted concrete is given in the table below. Let us do the calculation for a frequency of 1000 Hz.

Frequency	α
125	0.01
250	0.01
500	0.02
1000	0.02
2000	0.02
4000	0.03

The total surface area of the room is $96\,\text{m}^2$. The total absorption is

$$A = 0.02(96\,\text{m}^2) = 1.92 \text{ sabins}$$

Now let's cover the ceiling with acoustical tile. We shall assume the absorption coefficient for this material (at 1000 Hz) to be 0.85. The total absorption of the room now takes the form

$$A = 0.02(80\,\text{m}^2) + 0.85(16\,\text{m}^2)$$
$$= 15.2 \text{ sabins}$$

As we shall see in the next section, this *increase* in total absorption leads to a *decrease* in the reverberant noise of 9 db—a very significant change.

4 m
4 m
4 m
Concrete surfaces

Acoustical tile
ceiling added

Table 1 Absorption Coefficients for Various Materials

Material	125	250	500	1000	2000	4000
Air, per 100 cubic meters volume, relative humidity 50%	—	—	—	0.30	0.90	2.40
Concrete block, painted	0.01	0.05	0.06	0.07	0.09	0.08
Glass, window	0.35	0.25	0.18	0.12	0.07	0.04
Plaster, on lath	0.14	0.10	0.06	0.05	0.04	0.03
Plaster, on concrete block	0.12	0.09	0.07	0.05	0.05	0.04
Plaster, on brick	0.01	0.02	0.02	0.03	0.04	0.05
Plywood, paneling	0.28	0.22	0.17	0.09	0.10	0.11
Concrete block, unpainted	0.36	0.44	0.31	0.29	0.39	0.25
Drapery, lightweight	0.03	0.04	0.11	0.17	0.24	0.35
Drapery, heavyweight	0.14	0.35	0.55	0.72	0.70	0.65
Terrazzo floor	0.01	0.01	0.02	0.02	0.02	0.02
Marble floor	0.01	0.01	0.01	0.01	0.02	0.02
Wood floor	0.15	0.11	0.10	0.07	0.06	0.07
Carpet, on concrete	0.02	0.06	0.14	0.37	0.60	0.65
Carpet on pad	0.08	0.24	0.57	0.69	0.71	0.73
Carpet, indoor–outdoor	0.01	0.05	0.10	0.20	0.45	0.65
Acoustical tile: suspended	0.76	0.93	0.83	0.99	0.99	0.94
Upholstered furniture (fabric)	0.19	0.37	0.56	0.67	0.61	0.59
Upholstered furniture (leather)	0.44	0.54	0.60	0.62	0.58	0.50

SOUND WITHIN AN ENCLOSURE

Let us begin by considering a source of sound mounted on top of the Empire State Building. As an observer levitates away from the source, both the measured sound-intensity level and the perceived loudness decreases. When the distance from the source is doubled, the intensity decreases by a factor of 4 and the sound-intensity level drops by 6 db. A graph showing the decrease in sound-intensity level with distance from the source is shown on page 182. If our source is located at ground level, the sound-intensity level might not drop quite so rapidly since some of the sound energy would be reflected from the ground; however, the loudness still drops off rapidly with distance from the source.

Now let's bring our sound source indoors. If we here measure the sound-intensity level at different distances from the source, we find that the results differ significantly from the open-air case. A graph showing the sound-intensity level vs. distance within an enclosure is shown on page 182. It is clearly different from the previous graph. For small distances from the source (less than one meter), the *SL* decreases in the

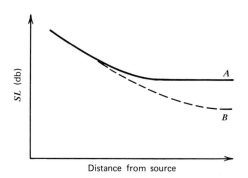

same way as it did in the open-air example. But then, at some distance
r, the sound-intensity level stops decreasing and assumes a constant
value for all greater distances. Why is this so? The answer lies in the
buildup of reflected sound.

In an enclosure the first sound to reach the ear of an observer is that
sound traveling directly from the source to the ear. This direct sound is
soon joined by sound reflected off one surface. Successive waves of
once-reflected sound move by the observer. Then sound that has
traveled a little further and has been reflected off of two surfaces reaches
the ear of the observer. Again, successive waves of twice-reflected sound
impinge upon the observer's ear. This same line of reasoning can be
extended to multiple-reflected sound waves. Of course, with each
reflection some energy is absorbed by the surface so that the increments
of energy arriving at the observer decrease with each added reflection.
The buildup of reflected sound is illustrated on opposite page.

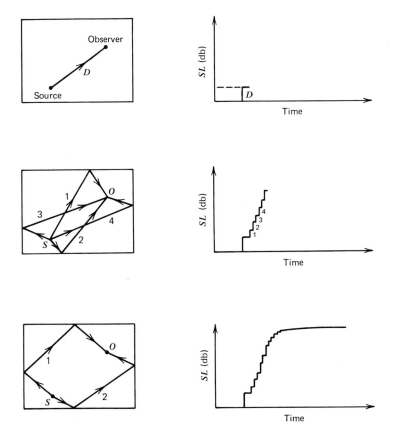

A steady state is reached in which the acoustical energy emitted by the source equals the acoustical energy absorbed by the environment. The steady state sound-intensity level depends on the total absorption of the environment. Curve A in the graph on page 182 shows a high steady state *SL*. If the surfaces of the environment are treated with an absorbing material, the steady state level decreases as shown in curve *B*.

Suppose the sound source is impulsive rather than sustained? We can deduce the behavior of sound under this circumstance following the line of reasoning we used before. After the retort of a gunshot, the first sound to reach the detector is the direct sound followed successively by once-reflected sound waves, twice-reflected sound waves, etc. The process is represented below. The persistence of the sound after the direct sound is referred to as *reverberation*. The reverberant sound slowly decays as energy is absorbed by the surfaces. The more absorbing the surface, the faster is the decay. The decay of sound follows a curve having a form similar to the buildup curve in the case of a sustained sound.

In the previous section, the effect of covering a ceiling with acoustical tile was determined for a concrete $4 \, \text{m} \times 4 \, \text{m} \times 4 \, \text{m}$ enclosure. We can now better understand what this means. The effect of adding acoustical tile is the reduction in the steady state sound-intensity level or the reverberant sound level. In the example considered the total absorption increased from 1.92 to 15.2 sabins. From these total absorption figures we can calculate the reduction in reverberant sound level. Noise reduction, NR, is defined below:

$$NR = 10 \log \frac{A_2}{A_1}$$

where the ratio A_2/A_1 essentially compares two total absorptions. The unit of NR is decibels (db). (**Note.** This definition is mathematically equivalent to the definition of sound-intensity level given in Chapter 3.) From the above definition the noise reduction can be calculated:

$$NR = 10 \log \frac{15.2}{1.92}$$

$$= 10 \log 7.92$$
$$= 10(0.899)$$
$$= 8.99 \, \text{db} \simeq 9 \, \text{db}$$

If the enclosure in question were an office and we reduced the noise level by 9 db, we would have made a significant improvement to the peace of mind of the office core.

There is one enclosure with which most people are familiar: the bathroom shower stall. In this unusual enclosure, the most ordinary singer is transformed into a veritable Caruso or Ella Fitzgerald. How does this happen? First, the surfaces of a shower stall are hard and smooth; therefore, they are highly reflective. The reverberant sound level is high, which leads to the allusion of a big, full-bodied voice. Second, singing in a shower stall is like singing in an organ pipe; that is, the air in the shower stall has natural vibrational modes that are excited by a singing voice. The shower stall literally resonates like an organ pipe.

If an enclosure has dimensions $l_x \times l_y \times l_z$, the resonant frequencies are given by

$$f_{xyz} = \tfrac{1}{2} v \sqrt{\left(\frac{n_x}{l_x}\right)^2 + \left(\frac{n_y}{l_y}\right)^2 + \left(\frac{n_z}{l_z}\right)^2}$$

where

$$v = \text{the speed of sound} \simeq 345 \text{ m/sec}$$
$$n_x, n_y, \text{ and } n_z \text{ are integers } 0, 1, 2, 3, 4, \ldots$$

A shower stall in my home measures as follows:

$$l_x = 0.74 \text{ m}$$
$$l_y = 1.19 \text{ m}$$
$$l_z = 2.13 \text{ m}$$

The lowest resonance frequency is

$$f_{001} = \tfrac{1}{2}(345) \sqrt{\left(\frac{0}{0.74}\right)^2 + \left(\frac{0}{1.19}\right)^2 + \left(\frac{1}{2.13}\right)^2}$$

$$f_{001} = 172.5 \sqrt{\left(\frac{1}{2.13}\right)^2}$$

$$= \frac{172.5}{2.13} = 81 \text{ Hz}$$

Other low resonant frequencies are:

$f_{001} = 81$ Hz	$f_{002} = 162$ Hz	$f_{003} = 243$ Hz
$f_{010} = 145$ Hz	$f_{020} = 290$ Hz	$f_{030} = 435$ Hz
$f_{100} = 235$ Hz	$f_{200} = 470$ Hz	$f_{300} = 705$ Hz
$f_{011} = 167$ Hz	$f_{022} = 334$ Hz	$f_{033} = 501$ Hz

If you calculate some other frequencies—for example, f_{101}, f_{110}, and f_{111}—and arrange them in ascending order, you will find there are many, many resonant frequencies that come close together. The singing voice excites many of these and the shower stall is prompted to sing along. In short, the bather receives full support from this acoustic environment.

**QUANTITATIVE
ACOUSTICAL
PARAMETERS**

Several acoustical parameters serve to characterize a concert hall or any other enclosure. These parameters might be called the vital statistics of an acoustical environment.

Reverberation Time

Perhaps the most important characteristic of any enclosure is its reverberation time. We have already seen that following the direct sound from an impulsive source, one reflected wave after another impinges on the observer. The reverberation time is a quantitative measure of how rapidly the reflected sound decays. The time required for the intensity of an impulsive sound to decrease by a factor of one million is defined as the reverberation time. Put another way, the reverberation time, R_T, is the time required for the sound-intensity level to decrease by 60 db.

What determines the value of R_T? Professor W. C. Sabine established a relationship that equates the reverberation time to other measurable room parameters. This relationship is

$$R_T = \frac{0.16\,V}{A + \alpha_a V}$$

where

V = the volume of the enclosure (in cubic meters)

A = the total absorption (in square meter sabins)

α_a = the absorption coefficient of air

R_T = reverberation time (in seconds)

Let us use this relationship to calculate the reverberation time of the $4\,\text{m} \times 4\,\text{m} \times 4\,\text{m}$ room considered in a previous section. We shall assume a humidity of 50 percent and use the absorption coefficient of air for a frequency of 2000 Hz. (**Note.** Air absorbs very little acoustic energy below frequencies of 1500 Hz.)

The appropriate absorption coefficient is found in Table 1. The calculated total absorption (with acoustical tile covering the ceiling) was 15.2 sabins.

$$R_T = 0.16 \frac{64\,\text{m}^3}{15.2 + 0.9\frac{64}{100}}*$$

$$= 0.16 \frac{64}{15.2 + 0.6}$$

$$= 0.16(4.05) = 0.65 \text{ sec}$$

From this example it can be noted that air absorption is only significant in large enclosures.

An audience increases the total absorption and thereby decreases the reverberation time. The absorbing properties of one person is approximately equivalent to $0.5\,\text{m}^2$ of open window. A guest conductor may notice a significant difference between the sound of the orchestra during rehearsal and later when the concert hall is full. Some reverberation times of well-known concert halls (occupied) are given opposite.

*In Table 1, the absorption coefficient of air is given per 100 cubic meters; therefore, the volume in the denominator is divided by 100.

Boston Symphony Hall	1.8 sec
Cleveland Severance Hall	1.7 sec
New York Carnegie Hall	1.7 sec
Philadelphia Academy of Music	1.4 sec
St. Louis Powell Hall	2.2 sec

(All the reverberation times above are measured at mid-range frequencies of 500 to 1000 Hz.)

The calculation of R_T using the formula given above is simple—perhaps deceivingly simple. The formula *does* give results agreeing with experiment for many enclosures; namely, those in which the sound energy becomes distributed uniformly throughout the enclosure. In other situations, however, the formula fails. One example is the situation where an enclosure has a very irregular shape. Another is where the surfaces are either highly absorbing or where the absorbing surfaces are concentrated in one particular area. One should further realize that the measurement of absorption coefficients is tricky: the results obtained may depend on the methods used to make the measurements. For this reason there is disagreement in the literature on the values of absorption coefficients.

Initial-Time-Delay Gap

A colleague of mine is blind. Yet, he can walk into a strange room and tell whether the room is large or small. Undoubtedly he is able to do this by taking note of the initial-time-delay gap. The interval of time between the perception of the direct sound and the perception of the first reflected sound is the initial-time-delay gap. We shall symbolize this quantity t_I and show it in the figure below.

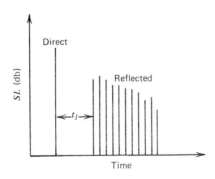

Consider the paths followed by the direct sound and the reflected sound, as illustrated below. The times required for the direct sound and

reflected sound to arrive at the observer are

$$t_D = \frac{R_D}{v} \qquad (v = \text{speed of sound})$$

$$t_R = \frac{R_1 + R_2}{v}$$

respectively. The initial-time-delay gap is the difference between t_D and t_R. That is,

$$t_I = t_R - t_D = \frac{R_1 + R_2 - R_D}{v}$$

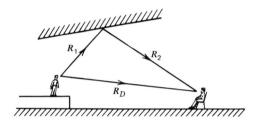

If we express distances in meters and if we assume $v = 345$ m/sec (room temperature), then

$$t_I = \frac{R_1 + R_2 - R_D}{345}$$

It should be obvious that t_I is not the same for all seats in a concert hall. Seats close to the orchestra will have a larger t_I than those seats in the balcony. Some values are given below. It is generally agreed that t_I should be less than 0.030 sec for good listening conditions.

Concert Hall	t_I Mid-Main Floor	t_I Mid-Balcony
Boston Symphony Hall	0.015 sec	0.007 sec
Cleveland	0.020 sec	0.013 sec
New York Carnegie Hall	0.023 sec	0.016 sec
Philadelphia	0.019 sec	0.010 sec

The Ratio

$$\frac{I_{\text{direct sound}}}{I_{\text{reverberant sound}}}$$

The loudness of a tone sounded in an open field depends solely on the energy output of the source and the distance between the source and the observer. In an enclosure, on the other hand, the loudness of a perceived tone depends both on the energy output of the source plus the energy of the reverberant sound. A useful measure is the ratio of the intensity of direct sound to the intensity of the reverberant sound.

ACOUSTICIANS INTERPRET-ATION OF THE CRITICS TERMINOLOGY

Earlier we quoted some music critics. These critics employ a common language, using words such as reverberant, intimate, warm, full, balance, brilliance, clarity, detailed, and blend. These terms have a subjective quality about them; however, we can now provide an acoustical interpretation. In essence, we can translate the language of the critic into the language of the acoustician.

Intimacy

Intimacy, or its lack, is determined by the initial-time-delay gap. If the initial-time-delay gap is large, a listener has the feeling of being in a cavernous, somewhat hostile, environment. Conversely, a small t_I leads to a sense of intimacy and at-homeness.

Psychoacoustical experiments have been performed in which tape recorders were used to simulate different t_I's. Critical listeners found initial-time-delay gaps of 0.015 to 0.020 sec disturbing. A t_I of 0.070 disturbed 80 percent of the subjects. On the basis of these experiments, it can be concluded that an initial-time-delay gap greater than 0.020 sec detracts from the quality of the music and a feeling of intimacy cannot be achieved if $t_I > 0.070$ sec.

Fullness of Tone and Clarity (Reverberant)

In a reverberant enclosure, a tone remains audible after the source ceases. The period of audibility depends on the reverberation time. Clearly, if the reverberation time is too long, a tone can become garbled with the reverberant sound of the preceding tones. On the other hand, if the reverberation time is too short, each tone stands alone and isolated from its neighbors in a musical sequence.

Fullness of tone depends on the reverberation time and more directly upon the ratio

$$\frac{\text{Loudness of reverberant sound}}{\text{Loudness of direct sound}}$$

In order for an acoustical environment to provide fullness of tone this ratio must be large. The larger this ratio, the greater the fullness of tone.

When a critic speaks of clarity, he is referring to the degree to which individual tones stand apart one from another. Once again, the acoustical parameters determining whether an environment provides clarity are the reverberation time and the above ratio. This time, however, both T_R and the ratio must be small. Fullness of tone and clarity are acoustical counterparts. An enclosure with fullness of tone lacks clarity and vice versa.

Warmth and Brilliance

A particular surface may absorb low frequencies more efficiently than it does high frequencies. Therefore, the reverberation time can be small for low frequencies and larger for high frequencies. The converse can also be true.

An enclosure is said to be acoustically warm if bass tones are full relative to mid-range frequencies. More specifically, if T_R for frequencies below 250 Hz is greater than T_R for the mid-range frequencies (500–1000 Hz), such an enclosure is judged as warm. Most highly regarded concert halls have a warm sound.

Brilliance is that quality resulting from the perception of high frequencies. For an environment to display brilliance, the treble frequencies must be prominent and they must decay slowly. In other words, T_R for high frequencies must be larger than T_R for mid-range frequencies.

These terms and more like them are becoming better understood in terms of basic acoustical concepts. With this understanding, the possibility of being able to design acoustical environments with predetermined properties becomes a goal more nearly attainable.

THE INTER-DEPENDENCE OF ARCHITECTURE AND MUSIC

Music written in a certain period share stylistic similarities. These similarities enable historians to identify stylistic musical types: Baroque, Classical, and Romantic. While musical form was undergoing change, architectural form was also changing. So close, in fact, is the relation between architecture and music that questions can be raised: did composers create musical forms with a particular acoustical environment in mind? Did architects note the direction being taken by composers and design structures to accommodate their music? Does a style of music require a particular acoustical environment?

Baroque Period

The Baroque period stretches from about 1600 to the death of J. S. Bach in 1750. Baroque composers were employed by a church, a municipality, a court, or perhaps an opera house. They wrote music for specific occasions—a religious celebration, a wedding, or a festival. These occasions took place in particular places: perhaps a ballroom, a theater, or a church. These places shared common characteristics. They were small, generally rectangular, and had hard reflecting walls.

Such qualities imply certain perceptive attributes. These environments were intimate with a short reverberation time. The music played in this environment had great clarity and little fullness of tone. Today, Baroque music is enjoyed the most in a small, intimate environment with $T_R < 1.5$ second.

It might be pointed out that there is more variation in Baroque sacred music. This is because there was a greater variation among churches and cathedrals. Some were very large and highly reverberant. Consequently, Baroque organ music typically benefits from a long reverberation time. time.

Classical Period The classical period centers on the unparalleled achievements of the Viennese school with its masters Haydn, Mozart, Beethoven, and Schubert. This period extended from approximately 1775 to 1825. The modern orchestra was established by the classicists. A smaller number of instrumentalists was characteristic of the classical era's orchestras—perhaps as many as 40 players. Yet they were similar to today's orchestras in form, containing string, woodwind, brass, and percussion sections.

The early concert halls were considerably smaller than their modern-day counterparts. Capacities of 300 to 400 people were typical. When these concert halls were fully occupied, they had reverberation times shorter than 1.5 second. Larger structures were built throughout the nineteenth century with corresponding increases in reverberation time to the 1.5 to 1.8 second range.

Classical music is best performed today in concert halls with reverberation times in the range of 1.5 to 1.7 seconds. These times are consistent with the reverberation times of the concert halls in Vienna, Leipzig, and Oxford used during the late classical period.

Romantic Period Romantic music is a much more personal and emotional expression of the composer. Fullness of tone and longer reverberation times are required to bring out the expressive intent of composers such as Brahms, Tchaikovsky, Wagner, and Debussy. This period, 1825 to 1900, saw larger concert halls coming into use with reverberation times reaching the duration of two seconds.

Today Romantic music is preferred in an acoustic environment with a reverberation time in the 1.9 to 2.2 second range.

REMEDIAL ACOUSTICS The acoustical problems of a poorly designed auditorium generally fall into four categories.

1. Background Noise Interferes with the Sound Program

We live in a noisy society. External noise such as traffic, the sirens of emergency vehicles, and airplanes can be annoyingly distractive. Mechanical noises originating within a structure can also be disruptive; for example, ventilating systems, air conditioning equipment, and heating equipment are potentially troublesome. There are many strategies for dealing with ambient noise. Air ducts can be lined with sound-absorbing materials, bends can be introduced into duct systems, ducts can be decoupled from walls or floors at their entry point, and the noise can be diffused at air supply and return points. Motors can be shock mounted to reduce vibrations being transmitted through floor and wall materials.

2. Reverberation Time is Excessive

Excessive is a relative term. An auditorium acceptable for music may not be adequate for lectures. The following graph gives reverberation times considered optimum for various functions. Excessive reverberation results from either an enclosed volume that is too large—frequently the ceiling height is too large—or one whose total absorption is too small. Ceilings can be lowered and absorbing materials can be added to the surfaces. Carpets, draperies, and upholstered seats are all options for consideration.

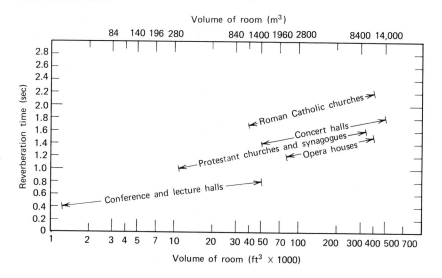

Decorative or geometric designs on a wall can also add to the total absorption. Protrusions from a wall traps sound energy. Consider, for example, the figure below. Such a configuration not only can trap acoustic energy, but can do so in a predetermined fashion. The determining factor is the interprotrusion spacing, x. Wavelengths that are approximately equal to x will be diffracted by the protrusions and will be absorbed more efficiently than wavelengths for which $\lambda \ll x$.

3. Sound Program Loudness is Inadequate

The surfaces in the vicinity of the sound source are critical. A stage should have reflecting surfaces for both the walls and ceiling. Elevating the stage can also be helpful. Reflecting panels suspended from the ceiling can distribute the sound more evenly and increase the acoustic energy arriving at the listener.

4. Echos, Hot Spots, and Dead Spots Are Troublesome

Echos are caused by reflecting surfaces far from the source such that there is a time delay between receipt of the direct sound and reflected sound. Time delays greater than 40 milliseconds can be troublesome. Acoustical treatments of the villianous surfaces can reduce the echo problems.

Hot spots and dead spots are areas where sound energy is focused or where sound energy is sparse. Once again, surfaces can be treated and reflecting panels can be employed to cope with this problem.

For Further Thought

1. Let us understand by means of an example the distinction between regular and diffuse reflection. Light can be collimated into a well-defined narrow beam. Imagine such a beam of light striking an apple pie sitting on a pedestal. The room is darkened. The pie can be seen from every position in the room. Why? Because reflection from the pie crust is diffuse and the light is scattered in all directions. Suppose the light beam was incident upon a very smooth mirror. Describe where the reflected light could be observed.

2. Why would you expect regular reflection of sound to be much more unusual than regular reflection of light? Acoustically speaking, is it good that regular reflection of sound is rare? Why?

3. At a frequency of 500 Hz, plywood paneling has an absorption coefficient of 0.17. What is the absorptivity of a piece of this paneling that measures $2 \, \text{m} \times 3 \, \text{m}$?

4. A room has dimensions of 6 m (width) × 8 m (length) × 3 m (height). The walls and ceiling are plaster on lath; the floor is wood. The room has one archway ($2 \, \text{m} \times 2.5 \, \text{m}$) and two windows (each window has dimensions of $1 \, \text{m} \times 2 \, \text{m}$). What is the total absorption of this room?

5. For the room described in question 4, determine the noise reduction that will result if the ceiling is covered with acoustical tile and a padded carpet is added to the floor.

6. A closet has dimensions of $2 \, \text{m} \times 1.5 \, \text{m} \times 3 \, \text{m}$. What are the three lowest resonant frequencies of this enclosure?

7. Again consider the room described in question 4 above. Calculate the reverberation time for this room. Ignore air absorption.

8. Some additions were made to the room described in question 4. These additions are described in question 5. Calculate the reverberation time for the room with the additions. Again, ignore air absorption.

9. A gymnasium has the following dimensions: width, 32.3 m; length, 47.5 m; height, 14.6 m. The walls are painted concrete block; the ceiling is a composite material, which we can assume is acoustically equivalent to plywood, and the floor is wood.

(a) What is the reverberation time of this enclosure?

(b) Estimate what the largest initial-time-delay gap would be.

(c) Occasionally, musical programs are held in this gymnasium. Comment on the acoustical characteristics of this enclosure.

13 Perception of Music

Music in our time is virtually ubiquitous. Whether we like it or not we eat to music, ride elevators to music, shop to music, get our teeth drilled to music, sit captive in airplanes to music, perform our bathroom functions to music, and navigate the highways and airways to music. In such a music-prone society, it is easy to become cavalier and "tune out" music in the same way we disregard annoying background noise.

SUBJECTIVE MUSIC AND OBJECTIVE ACOUSTICS

Listening to music is itself an art. Music is an art that conveys meaning and probes the limits of a person's emotions and intellect. The meaning is conveyed by melody, consonance, dissonance, rhythm, tempo, and dynamics (variations in loudness). The listening experience and the meaning it brings are very personalized.

While the listening experience is individualized and unique, the pressure disturbances and the psychological sensations induced by these stimuli are shared by every listener. In this dichotomy lies a methodological and philosophical morass. Many people seek an understanding of the musical experience based on the psychological sensations to which it gives rise. Others maintain that such an approach misses entirely the organizing and transforming action of the mind upon these sensations. The difference between these two points of view suggests a prevalent danger; namely, the tendency to assume that the physical stimuli and the psychological sensations they engender are in fact equivalent to and the essence of the musical experience.

In the analysis of musical sound perception, various physical parame-

ters immediately emerge as significant: for example, frequency, amplitude, waveform, and ratios of frequencies. A subjective response to these physical stimuli can be noted. Since the physical stimuli can be measured precisely and defined with exactness and since the rudimentary sensations provoked by these stimuli can be identified, some are tempted to "explain" the whole of the musical experience in terms of these stimuli and sensations. Next, in this "explanation" the emotive dimension of the musical experience becomes ephemeral and, in reality, only a figment. This line of reasoning, I believe, is fallacious.

In discussing the physics of musical sound, it is important to keep the limitations of physics sharply in focus. Physics can accurately measure frequency, but a person perceives pitch. The intensity of a sound wave can be measured, but the human auditory system perceives loudness and does so in a characteristically "human way." (See graph on page 42.) A complex tone can be analyzed into its harmonic components, but a listener perceives timbre. Moreover, music is not an isolated complex tone of a certain frequency and loudness; rather, it is a rhythmic succession of tones, each deriving its meaning from the context that all the tones create.

The perception of musical sound is a complicated process. Suppose, for example, that a complex tone reaching the ear can be represented by the audio spectrum (*a*) shown below. Such a tone is perceived as having

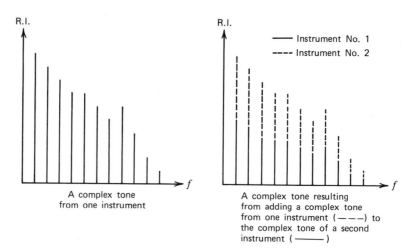

A complex tone
from one instrument

A complex tone resulting
from adding a complex tone
from one instrument (– – –) to
the complex tone of a second
instrument (——)

a characteristic timbre determined, we say, by the harmonic structure of the tone. Suppose, however, that the stimulus shown in graph (*a*) is in actual fact the consequence of *two* different instruments, each with their own harmonic structure, playing a tone of the same pitch. The audio

spectra of the two instruments [figure (*b*) above] combine to give a stimuli equivalent to the spectrum of the single tone illustrated in (*a*). Nevertheless, our auditory system perceives the spectrum shown in (*b*) as *two* separate tones while spectrum (*a*) is perceived as *one* tone! How our auditory system accomplishes this feat is a mystery.

There are many questions concerning musical perception for which answers are tenuous at best. The problem of consonance and dissonance is one of the most critical areas in the subject of musical perception. To this topic we now direct our attention.

CONSONANCE AND DISSONANCE

If two tones played together are judged by the average listener as pleasant, the tonal combination is referred to as a consonance. On the other hand, if the combination is judged harsh, jarring, or unpleasant, it is a dissonance. Clearly, these terms have reference to subjective judgments; however, as we shall see, the basis for these judgments is a shifting basis and therefore makes an objective analysis difficult.

Dissonance is important in music. A musical score made up entirely of consonant intervals would probably be judged bland or even insipid by the mature listener. Dissonance brings a restlessness to a movement that is resolved in consonance. In the same way that suspense and conflict bring tension to drama, dissonance brings tension to music. Without this tension, the relaxation brought by consonance would be empty of meaning.

Music has become increasingly dissonant. Why is this? It is because the subjective basis of dissonance varies with time. A tonal combination introduced in one age can strike the musical audiences of that generation as harsh and dischordant. But, with time, audiences get accustomed to it and that dissonant interval becomes more and more consonant. Consequently, a later generation of composers has to create new dissonances to produce the same degree of tension as their predecessors. Compare, for example, Mozart's *Eine Kleine Nachtmusik* composed in 1787 with Arnold Schoenberg's *Vorgefuhle* composed in 1909. Such a comparison reveals that the history of music has been characterized by the emancipation of the dissonant interval.

Musical tradition and the listener experience circumscribed by that tradition is one dimension of the consonance-dissonance problem. Musical training, another kind of experience, is another dimension. Still another dimension emerges from the fact that a dissonant interval is more acutely dissonant when played by two similar instruments than when it is played by two different instruments.

These considerations should serve to illustrate the subjective nature of consonance and dissonance. Now let us see what objective considerations can be brought to bear on the subject.

The Beat Theory of Helmholtz

A complex tone is characterized by its harmonic structure. When two complex tones are played together, the harmonics of each tone are present in the stimulus arriving at the ear of the listener. For some tonal combinations the harmonic frequencies match each other exactly. For others they do not.

(i) The Unison

Let us consider a perfect unison. The figure below shows that the harmonic frequencies of each tone match each other exactly. The unison is pleasant to the ear and is considered a consonant.

(ii) The Whole-Note Interval

The situation changes with the whole-note interval. The ratio of frequencies between whole notes (on the tempered scale) is 1.123. If two complex tones a whole note apart are played together, the sound is unpleasant and is classed as a dissonance. The harmonic structure of each tone is shown in the following figure. There is an obvious mismatch for all frequencies. Furthermore, the frequencies are close enough together so that discernable beats as well as harshness can result. The beat theory of Helmholtz maintains that it is the beats that our auditory system responds to negatively and as a result the whole-note interval is unpleasant.

(iii) The Perfect Fifth

If complex tones of frequencies f and $\frac{3}{2}f$ are played together, the interval is a musical fifth. The harmonics of these two tones, illustrated below, show some matching frequencies and some potentially "colliding" frequencies.

The third harmonic of f_2 is between the fourth and fifth harmonics of f_1. With the fifth, however there are enough frequencies that match or are separated by a margin wide enough to be troublefree. The interval is therefore "beat free," and is a consonance.

Other musical intervals can be analyzed in similar fashion. An indication of how dissonant a musical interval is can be judged by how far along the harmonic sequence one has to go before a matchup of harmonic frequencies occurs. Table 1 shows various musical intervals and identifies the first pair of matching harmonic frequencies. As one proceeds down the table the number of colliding harmonics increases and the degree of dissonance increases. In the beat theory, the former is considered the cause of the latter.

The preceding table can be summarized as follows. A musical interval can be expressed as the ratio of two frequencies; that is,

$$\frac{f_2}{f_1} = \frac{n}{m}$$

where for most intervals n and m are integers. In general, the smaller the numbers, the greater the degree of consonance. Or, in terms of the beat theory, the small numbers imply an absence of disturbing beats among the higher harmonics.

There are numerous difficulties with the beat theory. An example or two will serve to illustrate these difficulties. It is reasonable to assume that a change in timbre, that is, a change in the relative intensities of the harmonics, would bring an alteration in the intensity of the resulting beats. Yet, a change in timbre does not alter the judgment of consonance or dissonance. In other words, a consonance produced by violins is also a consonance when it is produced by French horns. Furthermore, consider a major third and a major second. The harmonics of the major third C_3 to E_3 and the major second C_4 to D_4 are shown on page 200.

Table 1 Musical Intervals

Interval	Frequencies	First Pair of Matching Harmonics
Unison	f_1 and $f_2 = f_1$	f_2 with f_1
Octave	f_1 and $f_2 = 2f_1$	f_2 with $2f_1$
Fifth	f_1 and $f_2 = \frac{3}{2}f_1$	$2f_2$ with $3f_1$
Fourth	f_1 and $f_2 = \frac{4}{3}f_1$	$3f_2$ with $4f_1$
Third (major)	f_1 and $f_2 = \frac{5}{4}f_1$	$4f_2$ with $5f_1$
Sixth (major)	f_1 and $f_2 = \frac{5}{3}f_1$	$3f_2$ with $5f_1$
Third (minor)	f_1 and $f_2 = \frac{6}{5}f_1$	$5f_2$ with $6f_1$
Sixth (minor)	f_1 and $f_2 = \frac{8}{5}f_1$	$5f_2$ with $8f_1$
Whole tone	f_1 and $f_2 = 1.123f_1$	No match up
Semitone	f_1 and $f_2 = 1.0595f_1$	No match up

The Major Third: C₃ to E₃		The Major Second: C₄ to D₄	
C₃	E₃	C₄	D₄
130.8 ⟍ 34		261.6 ⟍ 32	
97 ⟩ 164.8		230 ⟩ 293.7	
261.6 ⟨ 68		523.3 ⟨ 64	
194 ⟩ 329.6		459 ⟩ 587.3	
523.3		1046.5	
	659.3		1174.7
1046.5		2093.0	

Included in this table are the frequency differences between corresponding harmonics. The critical differences occur between the respective fundamentals and the second harmonics. As can be seen, they are essentially the same. Yet, the major third is considered a consonance while the major second is a dissonance.

The Critical Band Theory of Plomp

Let us begin by considering pure tones. Specifically, we shall consider a fifth, two major thirds, and a major second. The fifth C₄ to G₄ is a consonance. The major thirds C₄ to E₄ and C₃ to E₃ are interesting because the former is definitely a consonance while the latter is marginally a consonance or perhaps even a dissonance. The major second C₄ to D₄ is a dissonance.

The fifth C₄ to G₄ involves frequencies of 262 and 392 Hz (assuming the tempered scale). The center frequency, that is, the average frequency, is 327 Hz. The critical bandwidth at 327 Hz is 100 Hz. This means that any two tones within the range 277 to 377 Hz will have a harsh, rough quality. The frequencies 262 and 392 Hz do not fall within this range.

With a series of experiments on consonance and dissonance judgment, Plomp has concluded that two pure-tone frequencies falling outside the critical band are judged consonant. The figure on page 201 illustrates the case of the fifth as being consonant. These experiments further reveal that the *most* dissonant interval is the one for which the two pure-tone frequencies are separated by 25 percent of the critical band. In fact, frequencies differing in the range from 5 to 50 percent of their corresponding critical band are typically judged dissonant. Therefore, we are led to conclude that two pure tones whose frequency difference is less than 50 percent of the appropriate critical bandwidth is a dissonance.

For the major third C₄(262 Hz) to E₄(330 Hz), the center frequency is 296 Hz for which $\Delta f_{CB} = 95$ Hz. The major third C₃(131 Hz) to E₃(165 Hz) has a center frequency of 148 Hz with $\Delta f_{CB} = 90$ Hz. Finally, the major second C₄(262 Hz) to D₄(294 Hz) has a center frequency of 278 Hz with

$\Delta f_{CB} = 95$ Hz. The figure above illustrates each of these intervals. The major third C_4 to E_4 is a consonance since the frequency difference of the two pure tones is greater than $0.5\Delta f_{CB}$. One octave lower in frequency, however, the major third C_3 to E_3 is certainly less consonant in terms of the Plomp criterion and may even be considered a dissonant interval. The major second C_4 to D_4 is certainly a dissonance.

The experimental results obtained for pure tones can now be extended to complex tones. Each complex tone in a musical interval has a family of harmonics. If harmonics up to f_6 are considered for each complex tone, consonances will have a predominance of neighboring harmonics whose frequency difference is greater than $0.5\Delta f_{CB}$. A dissonance, on the other hand, will have neighboring harmonics whose frequency difference is less than $0.5\Delta f_{CB}$. (See figure on the next page.)

MUSICAL SCALES

Musical composition is based on musical scales. Western music divides the octave into 12 steps—semitone intervals—called the chromatic scale. If one starts at middle C on the piano and plays every note, white and black keys, to the C above middle C, one has played a chromatic scale. On the other hand, if only the white keys are played, one has played the C-major scale.

When the white keys are played from C to C, the familiar do-re-mi-fa-sol-la-ti-do sequence is heard. Looking at the piano, however, it is noted that there is no black key between E and F (mi and fa) or between B and C (ti and do). The separation between these two steps is a semitone while all other steps are separated by a whole tone. When we sing the do-re-mi sequence, we sing the sequence do, whole step, whole step, half step, whole step, whole step, whole step, half step. Sing the sequence and see if you can distinguish between the whole steps and the half steps.

There have been several ways devised to fill in the steps between the

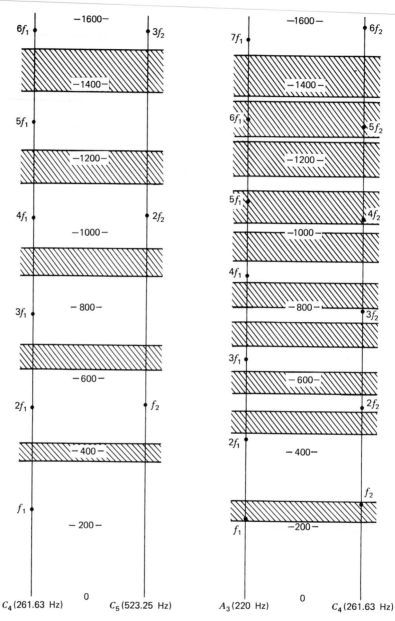

An octave (left) and a minor third (right) are illustrated. The shaded areas are centered on the average frequency between adjacent harmonics and they cover a frequency band equal to $0.5 \Delta f_{CB}$. Note that for the minor third f_1 and f_2, $5f_1$ and $4f_2$, $6f_1$ and $5f_2$ fall within a shaded area. For the octave, all harmonic frequencies are outside the shaded areas. By the Plomp criterion, the minor third A_3 to C_4 is more dissonant than the octave C_4 to C_5.

initial and terminal do's of the scale. Unfortunately, none of the scales that have been proposed are free of difficulties. For example, all fifths should have a frequency ratio of $\frac{3}{2}$, all fourths a ratio of $\frac{4}{3}$, and all major thirds a ratio of $\frac{5}{4}$. No scale fulfills this requirement exactly, but some scales come closer than others. Furthermore, each whole tone difference should be the same. Some scales have two different whole-tone intervals.

What is a whole-tone interval? We can answer this question by considering the consonant intervals. The ratio of frequencies sol/do $= \frac{3}{2}$. The ratio of frequencies fa/do $= \frac{4}{3}$. The interval between fa and sol is a whole tone. From the ratios above we can write

$$\frac{\text{sol/do}}{\text{fa/do}} = \frac{\text{sol}}{\text{fa}} = \frac{3/2}{4/3} = \frac{9}{8} = 1.125$$

Therefore,

$$\text{sol} = \frac{9}{8}\,\text{fa} = 1.125\,\text{fa}$$

From this we can conclude that in a whole-step interval the frequency is increased by the factor $\frac{9}{8}$ or decreased by the factor $\frac{8}{9}$.

(i) The Pythagorean Scale

The Pythagorean scale can be developed from the musical fifth. Let us begin with an arbitrary note of frequency represented by 1. From this frequency we shall ascend in steps of musical fifths. This step is the first one indicated below.

Step 1 From an arbitrary beginning point, ascend in musical fifths.

$$1 \qquad \frac{3}{2} \qquad \left(\frac{3}{2}\right)^2 \qquad \left(\frac{3}{2}\right)^3 \qquad \left(\frac{3}{2}\right)^4 \qquad \left(\frac{3}{2}\right)^5$$

or

$$1 \qquad \frac{3}{2} \qquad \frac{9}{4} \qquad \frac{27}{8} \qquad \frac{81}{16} \qquad \frac{243}{32}$$

Step 2 Bring the notes obtained in step 1 into the range of a single octave by descending from these notes in whole-octave steps.

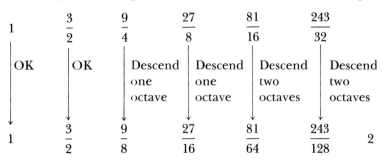

Step 3 Arrange the notes obtained in step 2 in ascending order.

$$1 \qquad \frac{9}{8} \qquad \frac{81}{64} \qquad \frac{3}{2} \qquad \frac{27}{16} \qquad \frac{243}{128} \qquad 2$$

Step 4 The above sequence of notes is missing one important note—the one corresponding to the musical fourth. This note can be obtained by descending from the beginning point by a musical fifth

$$\frac{2}{3} \longleftarrow 1$$

and then moving up one octave

$$\frac{2}{3} \longrightarrow \frac{4}{3}$$

These steps together give rise to the following sequence of

$$1 \qquad \frac{9}{8} \qquad \frac{81}{64} \qquad \frac{4}{3} \qquad \frac{3}{2} \qquad \frac{27}{16} \qquad \frac{243}{128} \qquad 2$$

do re mi fa sol la ti do

notes known as the Pythagorean scale. Let us examine the frequency interval between each step of the scale.

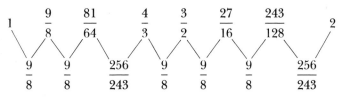

The ratio $\frac{9}{8}$ is a whole step and $\frac{256}{243}$ is a half step.

How adequate is this scale? Both the musical fifth and the fourth have exact ratios $\frac{3}{2}$ and $\frac{4}{3}$. Both are fine. The major third, however, is slightly off. The ratio of mi/do should be $\frac{5}{4} = 1.250$, but in the Pythagorean scale this ratio is $\frac{81}{64} = 1.265$. This means that mi is sharp relative to do. Furthermore, the minor third fa/re has a ratio $\frac{4}{3}/\frac{9}{8} = 32/27 = 1.185$. This is not equal to the desired ratio $\frac{6}{5} = 1.200$. Obviously, there are problems with the Pythagorean scale.

Another problem emerges if the above procedure is extended to generate the chromatic scale.

Step 1 From an arbitrary beginning point, ascend and descend in musical fifths.

$$\left(\frac{2}{3}\right)^6 \quad \left(\frac{2}{3}\right)^5 \quad \left(\frac{2}{3}\right)^4 \quad \left(\frac{2}{3}\right)^3 \quad \left(\frac{2}{3}\right)^2 \quad \frac{2}{3} \quad 1 \quad \frac{3}{2} \quad \left(\frac{3}{2}\right)^2 \quad \left(\frac{3}{2}\right)^3 \quad \left(\frac{3}{2}\right)^4 \quad \left(\frac{3}{2}\right)^5 \quad \left(\frac{3}{2}\right)^6$$

$$\downarrow \qquad \downarrow \qquad \downarrow \qquad \downarrow \qquad \downarrow \qquad \downarrow \qquad \qquad \downarrow \qquad \downarrow \qquad \downarrow \qquad \downarrow \qquad \downarrow$$

$$\frac{64}{729} \quad \frac{32}{243} \quad \frac{16}{81} \quad \frac{8}{27} \quad \frac{4}{9} \quad \frac{2}{3} \quad 1 \quad \frac{3}{2} \quad \frac{9}{4} \quad \frac{27}{8} \quad \frac{81}{16} \quad \frac{243}{32} \quad \frac{729}{64}$$

Step 2 Bring all the notes of step 1 within the range of a single octave by descending and ascending from these notes in whole octave steps.

$$\frac{64}{729} \quad \frac{32}{243} \quad \frac{16}{81} \quad \frac{8}{27} \quad \frac{4}{9} \quad \frac{2}{3} \quad 1 \quad \frac{3}{2} \quad \frac{9}{4} \quad \frac{27}{8} \quad \frac{81}{16} \quad \frac{243}{32} \quad \frac{729}{64}$$

$$\begin{array}{ccccccccccccc}
| & | & | & | & | & | & & | & | & | & | & | & | \\
+4 & +3 & +3 & +2 & +2 & +1 & & -1 & -1 & -2 & -2 & -3 \\
\text{octaves} & \text{octaves} & \text{octaves} & \text{octaves} & \text{octaves} & \text{octave} & \text{OK} & \text{OK} & \text{octave} & \text{octave} & \text{octaves} & \text{octaves} & \text{octaves}
\end{array}$$

$$\frac{1024}{729} \quad \frac{256}{243} \quad \frac{128}{81} \quad \frac{32}{27} \quad \frac{16}{9} \quad \frac{4}{3} \quad 1 \quad \frac{3}{2} \quad \frac{9}{8} \quad \frac{27}{16} \quad \frac{81}{64} \quad \frac{243}{128} \quad \frac{729}{512}$$

Step 3 Arrange the notes obtained in Step 2 in ascending order.

$$1 \quad \frac{256}{243} \quad \frac{9}{8} \quad \frac{32}{27} \quad \frac{81}{64} \quad \frac{4}{3} \quad \begin{array}{c}\frac{729}{512}\\ \diagup \quad \diagdown \\ \diagdown \quad \diagup \\ \frac{1024}{729}\end{array} \quad \frac{3}{2} \quad \frac{128}{81} \quad \frac{27}{16} \quad \frac{16}{9} \quad \frac{243}{128} \quad 2$$

	do♯		re♯			fa♯		sol♯		la♯		
do		re		mi	fa		sol		la		ti	do
	re♭		mi♭			sol♭		la♭		ti♭		

Step 3 identifies the problems. First, there are two semitone intervals. The intervals between do–do♯, re–re♯, mi–fa, sol–sol♯, la–la♯, and ti–do increase the frequency by a factor of $\frac{256}{243} = 1.053$. On the other hand, the intervals between do♯–re, re♯–mi, sol♯–la, and la♯–ti increase the frequency by a factor of $\frac{2187}{2048} = 1.068$. Second, there is an ambiguity concerning fa♯ (or sol♭) in that there are two different values for the same note: $\frac{729}{512}$ and $\frac{1024}{729}$.

These difficulties prompted the development of other scales.

(ii) The Just Scale

If we attempt to build a scale that maximizes the number of consonant intervals having *exact* frequency ratios, the just scale is the result. The major triad is a starting place. The harmony of Western music is built around the major triad, that is, frequencies in the ratios 4:5:6. The

notes do–mi–sol form a major triad as can be seen below:

do	mi	sol
f	$\dfrac{5}{4}f$	$\dfrac{3}{2}f$
$\dfrac{4}{4}f$	$\dfrac{5}{4}f$	$\dfrac{6}{4}f$
4	5	6

Step 1 Form a major triad:

do	mi	sol
f	$\dfrac{5}{4}f$	$\dfrac{3}{2}f$
4	5	6

Step 2 Form major triads from sol going up and from do going down:

		do	mi	sol		
		f	$\dfrac{5}{4}f$	$\dfrac{3}{2}f$		
fa′	la′	do		sol	ti	re′
$\dfrac{2}{3}f$	$\dfrac{5}{6}f$	f		$\dfrac{3}{2}f$	$\dfrac{15}{8}f$	$\dfrac{9}{4}f$
4	5	6		4	5	6

Step 3 Bring notes from step 2 within the range of a single octave by descending and ascending one octave.

fa′	la′	re′
$\dfrac{2}{3}f$	$\dfrac{5}{6}f$	$\dfrac{9}{4}f$
$+1$ octave	$+1$ octave	-1 octave
$\dfrac{4}{3}f$	$\dfrac{10}{6}f$	$\dfrac{9}{8}f$

Step 4 Arrange notes in ascending order

do	re	mi	fa	sol	la	ti	do
f	$\dfrac{9}{8}f$	$\dfrac{5}{4}f$	$\dfrac{4}{3}f$	$\dfrac{3}{2}f$	$\dfrac{10}{6}f$	$\dfrac{15}{8}f$	$2f$

The sequence of notes in step 4 is the just scale.

Let us examine first the interval between each note.

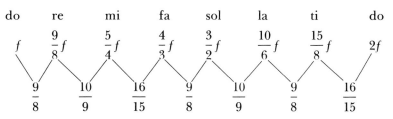

The first difficulty with this scale is that there are two whole tone intervals—one having a magnitude of $\frac{9}{8} = 1.125$ and another with a magnitude of $\frac{10}{9} = 1.111$. The semitone in this scale has a magnitude of $\frac{16}{15} = 1.067$. There are consonant difficulties also. The minor third fa/re $(\frac{4}{3}/\frac{9}{8} = \frac{32}{27})$ does not have the desired ratio $\frac{6}{5}$ nor does the fifth la/re $= \frac{10}{6}/\frac{9}{8} = \frac{40}{27}$ have the proper ratio $\frac{3}{2}$.

A chromatic scale can be constructed by a procedure analogous to that used for the Pythagorean scale. The ambiguities are more numerous for the just scale than they were for the Pythagorean scale. For example, la$^{\flat}$ is different from sol$^{\sharp}$; re$^{\flat}$ is different from do$^{\sharp}$; etc.

The just scale provides more pleasant harmonies than does the Pythagorean scale. However, another problem arises with the just scale. Suppose we have a piano tuned to the just scale. Let us assume the scale above represents the C-major scale. Suppose we now want to play the D-major scale, that is, *start* the do-re-mi sequence from D as shown below. Notice that the E's and A's are not the same. This means that it would be impossible for transposing instruments to play together properly.

		do	re	mi	fa	sol	la	ti	do
		C	D	E	F	G	A	B	C
C-major	1	$\frac{9}{8}$	$\frac{5}{4}$	$\frac{4}{3}$	$\frac{3}{2}$	$\frac{10}{6}$	$\frac{15}{8}$	2	
D-major	$\frac{9}{8}$	$\frac{81}{16}$	$\frac{45}{32}$	$\frac{3}{2}$	$\frac{27}{16}$	$\frac{15}{8}$	$\frac{135}{64}$	$\frac{9}{4}$	
	D	E	F$^{\sharp}$	G	A	B	C$^{\sharp}$	D	

All of these difficulties were clearly recognized by the seventeenth century. The resolution of these problems resulted in the scale to which we now turn.

(iii) The Tempered Scale

The Pythagorean scale is based on the fifth. The just scale is built on the third. Whenever exact fifths or thirds are fit into octaves there are bound to be misfits. The reason for this is the following: if you start from a particular note and ascend or descend in either fifths or thirds, one never

arrives at a note that is a whole number of octaves from the starting note. It is partly for this reason that there are problems with both the Pythagorean and just scales.

The tempered scale is an acoustical compromise designed to satisfy three incompatible requirements:

1. True intonation—that is, exact thirds, fourths, fifths, etc.
2. Complete freedom of modulation—that is, the freedom to make a musical transition from one key to another.
3. Convenience in the practical use in keyed instruments—that is, to make it possible for instruments tuned differently to play together.

In the compromise, the first of these requirements is sacrificed for the good of the latter two requirements.

In the tempered scale all semitone intervals are exactly the same whether it is do–do$^\#$, sol$^\#$–la, ti–do, or whatever. This implies that all whole-tone intervals are likewise equal. In order to accomplish this, the octave is divided into 12 equal intervals. We begin, then, with two operating demands:

1. That all semitone intervals are identical; that is,

$$\frac{f(\text{do}^\#)}{f(\text{do})} = \frac{f(\text{re})}{f(\text{do}^\#)} = \frac{f(\text{re}^\#)}{f(\text{re})} = \cdots = \frac{f(\text{do}')}{f(\text{ti})}$$

2. That the octave maintain its integrity; that is,

$$f(\text{do}') \equiv 2f(\text{do})$$

The first demand means that the ratio of all semitone intervals is the same. Let us call this ratio a. The two demands lead to the following frequency sequence for the chromatic scale.

do	f
do$^\#$	af
re	$a^2 f$
re$^\#$	$a^3 f$
mi	$a^4 f$
fa	$a^5 f$
fa$^\#$	$a^6 f$
sol	$a^7 f$
sol$^\#$	$a^8 f$
la	$a^9 f$
la$^\#$	$a^{10} f$
ti	$a^{11} f$
do'	$a^{12} f$

The ratio of any two semitones is the same:

$$\frac{\text{la}}{\text{sol}^\#} = \frac{a^9 f}{a^8 f} = a$$

$$\frac{\text{do}^\#}{\text{do}} = \frac{af}{f} = a$$

$$\frac{f(\text{fa})}{f(\text{mi})} = \frac{a^5 f}{a^4 f} = a$$

Furthermore, since

$$f(\text{do}') = a^{12} f(\text{do}),$$

we have

$$a^{12} = 2$$

or

$$a = \sqrt[12]{2}$$

The twelfth root of two is 1.059463 or 1.0595; therefore,

$$a = 1.0595$$

The tempered scale frequencies are given in Table 2.

The frequencies of the chromatic scale initiating on C_4 ($f = 261.6$ Hz) can be written down from Table 2. These frequencies are given in Table 3 where it is observed that once again, slight errors result because the value of a is rounded off.

Table 2

Note	Frequency
do	1.0000
do$^\#$	1.0595
re	1.1225
re$^\#$	1.1893
mi	1.2601
fa	1.3351
fa$^\#$	1.4145
sol	1.4988
sol $^\#$	1.5878
la	1.6823
la $^\#$	1.7824
ti	1.8885
do	2.0008[a]

[a]The discrepancy occurs because we rounded off the ratio from 1.059463 to 1.0595.

Table 3

C_4	261.6 Hz
$C_4\#$	$1.0595 \times 261.6 = 277.2$
D_4	$1.1225 \times 261.6 = 293.7$
$D_4\#$	$1.1893 \times 261.6 = 311.1$
E_4	$1.2601 \times 261.6 = 329.6$
F_4	$1.3351 \times 261.6 = 349.3$
$F_4\#$	$1.4145 \times 261.6 = 370.0$
G_4	$1.4987 \times 261.6 = 392.1$
$G_4\#$	$1.5878 \times 261.6 = 415.4$
A_4	$1.6823 \times 261.6 = 440.1$
$A_4\#$	$1.7824 \times 261.6 = 466.3$
B_4	$1.8885 \times 261.6 = 494.0$
C_5	$2.0008 \times 261.6 = 523.4$

Now let us examine the consonant intervals of the tempered scale. The interval sol to do is a musical fifth. From Table 3 this interval has a ratio of 1.4987. The correct ratio for the fifth is 1.5000; thus, the error introduced into the fifth by the demands of equal temperament is 0.087 percent. The fifth is flat. Other consonant intervals are shown on Table 4. It is clear from the results that equal temperament mistunes all intervals except the octave and mistunes thirds and sixths rather drastically. Consequently, there are intonation difficulties. In return for these difficulties, however, keyed instruments can play together freely, an orchestral score can be transposed downward a half-step for a hoarse tenor, and a composer can modulate from key to key with abandon.

Let us return to the intonation problem. Because the musical background of so many people is based upon or influenced by the piano, the piano scale is accepted as "natural." Is it "natural"? Not really. A good violin player or a vocalist plays or sings in an intonation different from equal temperament. In a string quartet the performers will play thirds and fourths the way they sound the best and that is not in equal temperament. A good orchestra does not play in equal temperament.

Table 4

Interval	Theoretical Ratio	Equal Temperament Ratio	Error
Octave	2.0000	2.0000	0%
Fifth	1.5000	1.4987	0.087% (flat)
Fourth	1.3333	1.3351	0.135% (sharp)
Major third	1.2500	1.2601	0.808% (sharp)
Minor third	1.200	1.1893	0.892% (flat)
Major sixth	1.6667	1.6823	0.936% (sharp)
Minor sixth	1.6000	1.5878	0.763% (flat)

Brass players tend to "lip up" or "lip down" certain notes and in so doing depart from equal temperament. If, for example, B_4 is being played as ti in the C-major scale, performers will tend to play it sharp in anticipation of do. If the same B_4 is being played as mi in the G-major scale, it will be played flat relative to the demand of equal temperament.

In equal temperament the scale is divided into 12 equal ratio intervals. Could some of the intonation difficulties be ameliorated if the scale was divided into more than or fewer than 12 intervals? From the early stages of the piano's development, there have been five black keys and seven

Table 5 Frequencies of Notes in the Tempered Scale

C_0	16.352		C_3	130.81		C_6	1046.5	
		17.324			138.59			1108.7
D_0	18.354		D_3	145.83		D_6	1174.7	
		19.445			155.56			1244.5
E_0	20.602		E_3	164.81		E_6	1318.5	
F_0	21.827		F_3	174.61		F_6	1396.9	
		23.125			185.00			1480.0
G_0	24.500		G_3	196.00		G_6	1568.0	
		25.957			207.65			1661.2
A_0	27.500		A_3	220.00		A_6	1760.0	
		29.135			233.08			1864.7
B_0	30.868		B_3	245.94		B_6	1975.5	
C_1	32.703		C_4	261.63		C_7	2093.0	
		34.648			277.18			2217.5
D_1	36.708		D_4	293.66		D_7	2349.3	
		38.891			311.13			2489.0
E_1	41.203		E_4	329.63		E_7	2637.0	
F_1	43.654		F_4	349.23		F_7	2793.8	
		46.249			369.99			2960.0
G_1	48.999		G_4	392.00		G_7	3136.0	
		51.913			415.30			3322.4
A_1	55.000		A_4	440.00		A_7	3520.0	
		58.270			466.16			3729.3
B_1	61.735		B_4	493.88		B_7	3951.1	
C_2	65.406		C_5	523.25		C_8	4186.0	
		69.296			554.37			4434.9
D_2	73.416		D_5	587.33		D_8	4698.6	
		77.782			622.25			4978.0
E_2	82.407		E_5	659.26		E_8	5274.0	
F_2	87.307		F_5	698.46		F_8	5587.7	
		92.499			739.99			5919.9
G_2	94.999		G_5	783.99		G_8	6271.9	
		103.83			830.61			6644.9
A_2	110.00		A_5	880.00		A_8	7040.0	
		116.54			932.33			7458.6
B_2	123.47		B_5	987.77		B_8	7902.1	

white keys to the scale. This combination seems to be particularly appropriate for the human hand. So perhaps we have been led by our hands rather than by our ears to subdivide the octave into the 12 intervals.

For Further Thought

1. Analyze (1) an octave and (2) a perfect fourth and in terms of the beat theory of Helmholtz show which harmonics might be expected to produce beats.

2. Consider the following complex tones that together form a musical interval.

 Interval 1: 130.8 and 164.8 Hz
 Interval 2: 261.6 and 329.6 Hz

 In terms of the beat theory, which interval would be the more consonant?

3. Perfect fourths have frequency ratios of $\frac{4}{3}$. The frequencies of three different perfect fourths are given below. In terms of the Plomp criterion, which of these fourths would be the most consonant? The least consonant?

 Fourth 1: 65.4 and 87.3 Hz
 Fourth 2: 261.6 and 349.2 Hz
 Fourth 3: 1046.5 and 1396.9 Hz

4. Examine the Pythagorean scale shown on page 204. How many perfect fourths are possible from the eight notes of this scale?

5. How many perfect major thirds are possible from the eight notes of the just scale shown on page 206?

6. Let us try to generate a scale by the following procedure: start at a frequency f, go up a perfect fifth, down a perfect fourth, up a fifth, down a fourth, etc. Identify the difficulties inherent in the scale resulting from this procedure.

7. Assume a musical score calls for the playing of a major third. It is to be played by a violin and a clarinet. Based on your understanding of consonance and dissonance, would you have the violin play the lower frequency or the higher frequency? Explain.

8. A new equally tempered scale is invented with 16 steps spanning the octave from f to $2f$. On this new scale, what is the frequency difference between the first note f and the fourth note of the scale?

Sound Reproduction

The goal of high-fidelity sound reproduction is twofold: first, to faithfully transmit and record a sound program signal. It matters not what the program is. It might be a symphony orchestra, an operatic production, country music, or any other musical form. The musical signal that is recorded should be in a form that is both compact and that permits repeated retrieval of the signal. Second, to faithfully reproduce or play back the recorded signal.

This goal is accomplished by a powerful and versatile means—electronics. The methods of electronics permit much more than the reproduction of sound with fidelity. Whether this is an advantage depends on one's auditory sensibilities. Through electronic means the sound can be modified in a variety of ways to please the individual listener. Now anyone who invests the requisite amount of money in a sound system can do what he or she wants with the controls of the system. However, the resulting acoustic signal may merely masquerade as the recorded signal. When a music program is recorded at an average sound-intensity level of 75 db and is played back at averages of either 60 or 90 db, the fidelity principle has been compromised. When electronic filters are used to emphasize the bass or the treble, thereby achieving what is referred to as "depth" on the one hand or "brilliance" on the other, fidelity has been lost. Likewise, when a flute is amplified up to the dimensions of a cathedral pipe organ a new sound has been created and fidelity abandoned. Again, something different is happening when a would-be bass "whispers" into a microphone held a centimeter from his lips and the resulting signal is amplified both at the recording and

213

playback stages. High fidelity—with integrity— avoids these kinds of perversions.

Several key events have contributed to the spectacular growth of the hi-fi industry. In the mid-1940s, FM radio, capable of providing static-free sound, was initiated. About the same time, Columbia Records marketed the long playing $33\frac{1}{3}$ rpm record, which not only improved the fidelity, but also increased the playing time of a record from a few minutes to 45 minutes or longer. In 1958 the first stereo record was released. Later, in 1961, the FCC authorized FM stations to transmit two signals that could be electronically separated into right and left channels. Stereo FM was the result. These developments provided a rich diversity of program material and stimulated the design of solid-state electronic circuits capable of exploiting these program sources to the limit.

After the purchase of a home and an automobile, the next largest investment an individual makes may well be a sound-reproduction system. Choosing the components to make up a sound system can be a frustrating experience. Component design is such a rapidly changing enterprise that today's state-of-the-art model can be tomorrow's antique. Audio zealots are among the more opinionated people of the world with the result that their advice is often more confusing than it is informative. Merchants are often motivated by commissions, profit margins, and stock surpluses rather than identifying the listening needs of a consumer and matching them to the appropriate sound system.

In Part VI we shall focus on sound reproduction systems that provide home entertainment. The components that comprise a sound system are identified. Component specifications are the objective criteria that enable quality judgments to be made. These specifications are defined in Chapter 14. Chapter 15 deals with the various configurations of these components and their placement within the room environment. Finally, we shall enter the arena in which the decision to buy is consummated. In Chapter 16 we shall make a modest comparison of hardware specifications with perceptual characteristics and discuss the role of perception more generally. A discussion of buying options and strategies concludes Part VI.

14 Sound System Components

Sound system components can be grouped into three classes: program source elements, program control elements, and program playback elements. We shall, of course, discuss all of them. Before doing so, however, we need to know some straightforward electromagnetic phenomena so that we can understand a most crucial element of sound reproduction systems—the transducer.

THE TRANSDUCER

Broadly speaking a transducer is a device that transforms a signal of one type into a signal of another type. An acoustic signal involves the systematic movement of a medium. It is a mechanical signal. When an acoustic signal is incident upon a microphone, the acoustic signal is transformed into an electrical signal. The microphone is a transducer. The resulting electrical signal is the input to various electronic circuits. Eventually that electrical signal may drive a stylus that cuts a wavy groove into a master disc. The cutting stylus "wiggles" implying that an electrical signal has been transformed back into a mechanical signal. Another transducer has come into play. A phonograph album has a wavy groove that the stylus on the tone arm of a turntable follows. In so doing, the mechanical motion of the stylus is converted into an electrical signal. The cartridge is a transducer. Later the electrical signal drives the cone of a speaker into mechanical motion, thereby initiating a disturbance in the air. A speaker is a transducer.

Transducers of the type considered above are based on electromagnetic principles. Electric and magnetic effects have been recognized since

215

the time of the early Greeks. Until 1820, however, electricity and magnetism were thought of as two distinct and unrelated phenomena. Then, in that year, it was discovered that an electric current has magnetic influences associated with it. In other words, a magnetic compass in the vicinity of an electric current is deflected. The figure below illustrates the manner in which compasses align themselves around a current-carrying conductor oriented perpendicular to the plane of the page. To describe this effect it is said that a current-carrying conductor produces a *magnetic field* in which the compasses align themselves.

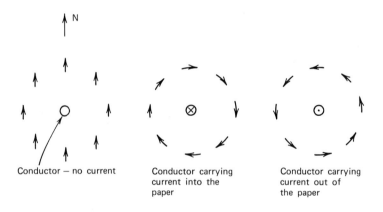

Conductor — no current Conductor carrying Conductor carrying
 current into the current out of
 paper the paper

In the immediate years that followed this 1820 discovery, the interrelationship between electricity and magnetism was the subject of intensive study. A very appealing idea developed that since an electrical effect—namely, an electrical current—gives rise to a magnetic field, a magnetic effect should give rise to an analogous electric effect. In other words, there should be a reciprocal relationship between electricity and magnetism.

Soon scientists became aware of a sort of reciprocity. The issue can be posed as a question: since an electrical current via a magnetic field can exert a force on a magnet (compass), should not a magnet via its magnetic field exert a force on a current-carrying conductor? This idea is a straightforward consequence of some of Newton's earlier work; therefore, it is not surprising that this reciprocity was discovered. The figure below shows the force exerted on a current-carrying conductor by a magnetic field.

But the large issue still remains: is there a completely reciprocal relationship between electricity and magnetism? Many approaches to this question were tried with no success. Finally, in 1831 Michael Faraday discovered the answer: yes, there is a reciprocal relationship. The answer was found not in a static magnetic field, but in a changing

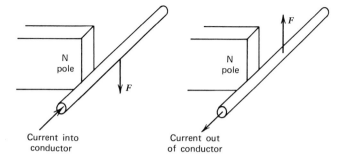

Current into conductor Current out of conductor

magnetic field. If a magnet is moved in the vicinity of a conductor, a current is induced in the conductor. By moving the magnet, the magnetic field surrounding the conductor is altered. A stationary magnet induces no current in the conductor. This effect is illustrated in the diagram below. It should be pointed out that it doesn't matter whether the magnet or the conductor is moving. What does matter is that the magnetic field in the neighborhood of a conductor is changing.

Magnet moving towards conductor — current induced

Magnet moving away from conductor — current induced in the opposite direction

The two principles of electromagnetism on which the transducers employed in sound systems are based can now be stated:

1. A force is exerted on a current-carrying conductor by a magnetic field.
2. A current is induced in a conductor by a changing magnetic field.

THE PHONOGRAPH CARTRIDGE Let us now examine one of the most important components of a sound system—the cartridge. The cartridge is a transducer that translates the mechanical wiggles of the stylus into an electrical signal. It is the cartridge that supplies the electrical signal that ultimately drives the loudspeakers. If the cartridge does not perform its task with impunity, it matters not how many thousands of dollars are invested in the sound

system. Shortcomings in the cartridge cannot be compensated for by subsequent components in the system. The conclusion to be drawn from these considerations is that a cartridge should be selected with care.

The requirements that a cartridge must satisfy are best understood if we understand how information is stored on a record disc. In simplest terms, a wavy groove is cut into the disc. As shown below, this groove has the characteristics of the waveforms we have studied earlier; namely, frequency, wavelength, and amplitude. With a stereo record each side of the groove is different, thereby providing two independent signals.

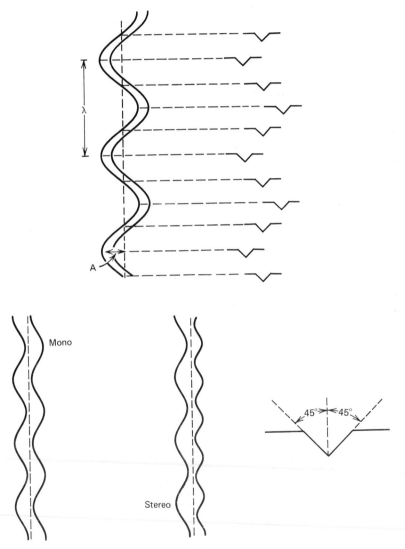

The stylus in the pickup cartridge tracks in this groove.

Magnification: 200 ×

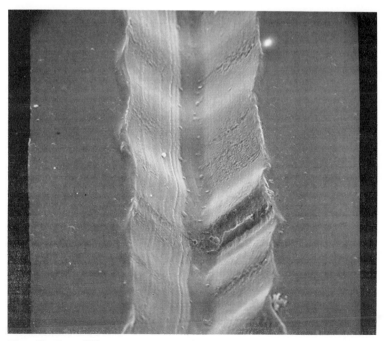

Magnification: 500 ×

Each side of the groove in a record disc makes an angle of 45° with the vertical. Clearly, the pick-up stylus cannot be pushed to the right by the left side of the groove while at the same time it is being pushed to the left by the right side of the groove. In response to the two undulating sides of the impressed groove, the stylus has both lateral motion (right-left motion) and vertical motion (up-down motion). The resultant motion of the stylus contains both right channel and left channel information that must be decoded by the phono cartridge.

There is another problem that arises because of the way the mechanical wiggles are transformed into an electrical signal. Earlier it was pointed out that a magnet moving in the vicinity of a conductor induces a current in that conductor. The faster the relative motion between the conductor and magnet, the greater is the induced current. The speed with which the stylus moves depends both on the rotational speed of the disc and the amplitude of the impressed signals. However, the stylus must move faster in tracing out a high-frequency curve (curve *B* in the diagram below) than it does a low-frequency curve (curve *A*) even though their amplitudes are the same. As a consequence, a low-frequency curve must have a correspondingly larger amplitude than its high-frequency counterpart if it is to induce an electrical signal of the same magnitude and hence give rise to an audio signal of the same intensity. In other words, curves *C* and *D* might produce sound of the same intensity.

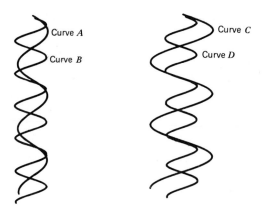

In order to achieve the desired dynamic range, compensatory steps must be taken to avoid overcutting those grooves corresponding to high-intensity low-frequency musical passages. Overcutting can cause one groove to cross over into the adjacent groove. Below a certain frequency called the turnover frequency, bass frequencies are progressively attenuated. This process is called *bass equalization.*

During the playback, the bass frequencies must be progressively

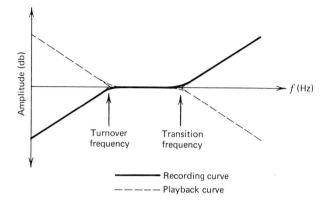

accentuated so that the final output has the same signal characteristics as existed before the equalization process.

Special considerations must also be taken in the recording of high-frequency signals. Very small groove amplitudes in the high-frequency range lead to intense audio signals. In fact, the amplitudes can be so small that discrepancies in groove shape and dust particles in the groove itself can significantly alter the signal during playback. To compensate for these problems high frequencies are accentuated during the recording stage. For frequencies higher than the *transition* frequency, the treble frequencies are progressively boosted thereby improving the signal-to-noise ratio. This process is called *preemphasis*. During playback, the treble frequencies above the transition frequency must be progressively attenuated.

The recording characteristic curve showing bass attenuation and treble preemphasis is shown above. Shown on the same graph is the playback equalization curve.

A phono cartridge must extract the information on a disc contained in the form of an undulating groove and transform it into an electrical signal. To achieve this, the stylus must track the groove—both sides of the groove—accurately. The contact force between the stylus and the groove should be small to minimize disc wear (and to prolong the life of the stylus) yet large enough so that the stylus does not inadvertently track on dust particles. Accurate trackability implies a sensitive responsiveness to the impulsive undulations of the record groove. For good response the mass of the stylus and other moving parts must be kept very small. Finally, the cartridge must deliver two clean signals where each signal is uninfluenced by the presence of the other signal and where each signal is strong enough for subsequent electronic stages to work with effectively. The quality specifications of a cartridge will reflect some of the considerations of this paragraph.

There are two types of styli commonly employed: one with a circular cross section and another with an elliptical cross section. Both have

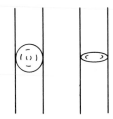

advantages and disadvantages. A stereo stylus has a tip radius on the order of 0.012 mm. The contact area between the stylus and the record groove is very small, which means that the pressures exerted by the stylus on the record groove can be very large even for small tracking forces. The elliptical stylus has a smaller contact area, which means that the contact pressure is greater than it is for a conical stylus. This means greater disc wear. In fact, an elliptical stylus tracking at 1 gram will produce as much wear as a conical stylus tracking at 2 grams. On the other hand, an elliptical stylus with its smaller contact area can negotiate the abrupt undulations somewhat better leading, in the opinion of some, to a "cleaner" sound. An elliptical stylus is more difficult to manufacture and mount. (A slightly rotated elliptical stylus can cause excessive groove wear.) For this reason, elliptical styli are more expensive.

There are several types of cartridges used for high-fidelity sound reproduction. The moving magnet (MM) and the moving coil (MC) cartridges are closely related. In the MM cartridge, a tiny permanent magnet is mounted on the shank of the stylus. The magnet is moved by the motion of the stylus, thereby inducing a current in neighboring conductors. In a MC cartridge, the magnet is fixed in a location and coiled conductors move in response to the stylus. The signal output of a MC cartridge is so small that an amplifier with high-sensitivity inputs must be used. Some of the most respected cartridges on the market today are the moving magnet type.

The operation of the ceramic cartridge is based on the piezo-electric effect. Some crystalline materials generate an electrical signal when forces are applied to them. This is called the piezo-electric effect. The motion of the stylus subjects the piezo-electric element to twists and stresses: electrical signals result from the applied forces. Although these cartridges are rather common in low priced systems, they do not meet the demands of the discriminating audiophile.

There are other cartridge types on the market, but they are not commonly used.

The following specifications are applicable to phono cartridges. In some cases the specification is also applicable to other components, but we will fully describe it only once and then merely refer to it when other components are discussed.

1. Frequency Response (in Hertz)

The lowest and highest frequencies to which the cartridge will respond define the response range. Generally, the wider the range the better. In other words, a frequency response of 20 to 20,000 Hz is better than one of 30 to 18,000 Hz. A response of at least 40 to 15,000 Hz is required to provide apparent fidelity; however, some musical types make special demands on the frequency response. For example, an organ aficionado probably demands a lower limit of 32 Hz on the response range.

The frequency range that a component will reproduce or respond to is crucial; however, without the next specification, frequency response is rather meaningless.

2. Uniformity of Response (in Decibels)

A frequency response of 20 to 20,000 Hz tells nothing of how uniformly the component responds across this range. It may boom out at 12,000 Hz and barely whisper at 16,000 Hz. Ideally it is desirable to have a flat response over the frequency response range. In actuality, a response uniform within 1 to 2 db is adequate for apparent fidelity.

A reputable frequency response specification includes the uniformity of response. For example, the specification is given as

$$20\text{–}20{,}000 \text{ Hz} \pm 2 \text{ db}$$

where the ± 2 db indicates that the response is uniform within limits of ± 2 db. Sometimes this specification is presented graphically.

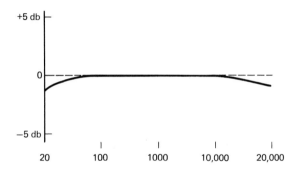

3. Output (in Millivolts, mV)

The cartridge generates an electrical voltage from the mechanical wiggles. The magnitude of the voltage depends on the relative speed between stylus and record. Therefore, this specification is given for a particular frequency and a particular stylus-to-groove speed. The outputs of various cartridges differ, but it should be about 2.0 mV or greater. (The cartridge output should be greater than the sensitivity of the amplifier's phono input.)

This specification can be expressed as follows:

<div align="center">

Output at 3.54 cm/sec

and 1000 Hz 3.4 mV

</div>

4. Channel Separation (in Decibels)

Each groove wall carries a signal. The cartridge must maintain the distinct individuality of these two signals. This specification indicates

how successfully the cartridge maintains the separation. A channel separation of 30 db means that the unwanted signal in a particular channel is 30 db below the wanted signal. Recall that the decibel unit compares two intensities or powers. Since power is proportional to the square of the voltage, we can write

$$\text{Channel separation (in db)} = 10 \log \frac{I}{I_{\text{ref}}}$$

$$= 10 \log \frac{V^2}{V_{\text{ref}}^2}$$

$$= 20 \log \frac{V}{V_{\text{ref}}}$$

Therefore, a channel separation of 30 db means that the wanted voltage is 32 times greater than the unwanted voltage.

The larger the channel separation, the better. It certainly should be greater than 20 db at 1000 Hz.

5. Tracking Force (in Grams)

Tracking force is the downward force that is required to make a cartridge stay in the record groove and track it. Tracking force should be neither too large or too small. A large tracking force accentuates disc wear. If the force is too small, the stylus will not track properly and severe distortions will result.

Tracking force is given in grams—a misnomer. Typical tracking force specifications range from 1 to 4 grams. (Monaural units have somewhat higher ranges.) Most better units track successfully at 1 gram.

6. Stylus Type

It will generally be indicated whether the stylus is conical or elliptical.

7. Load Impedance (in Ohms)

The output of a cartridge must be matched to the input of the amplifier. A MM cartridge has a load impedance in the range 47,000 to 50,000 ohms, which matches the inputs of most amplifiers. A MC cartridge has a low impedance—2 to 20 ohms. For a MC cartridge, most amplifiers must have special phono input.

HIGH-FIDELITY PROGRAM SOURCES

A sound reproduction system requires a program source. The primary high-fidelity sources are the phonograph turntable, the tape deck, and the FM tuner. We shall consider each of these.

The Phono Turntable or Changer*

A record disc has impregnated upon it a curve. Basically, a turntable provides a means for (1) turning the record at a very steady RPM, and (2) holding a transducer—the cartridge—at the proper position with respect to the record groove. A changer adds to this a means for automatically changing the record disc.

Now, what is important? First, a mechanism must rotate the disc at a constant rotational speed very near the RPM called for by the record disc. A $33\frac{1}{3}$ RPM record rotated at $32\frac{1}{2}$ RPM will sound flat while at 34 RPM it will sound sharp. More important, the rotational speed cannot vary. *Wow* and *flutter* are slow and rapid unevenness of pitch. These maladies result from variations in the rotational speed.

An electric motor via mechanical linkages drives the turntable platter. The drive system must be engineered so that is is free of vibrations. *Rumble* is unwanted mechanical noise caused by the turntable mechanism. The electrical motor should not produce any magnetic fields that can get picked up by the cartridge and cause the noise referred to as *hum.*

There are three types of drive systems. The *rim drive* mechanism transfers the power from the motor shaft to the turn-table platter via one or more idler wheels. The idler wheel, in contact with the platter, is rubber rimmed and drives the platter by friction. The chief advantage of the rim drive turntable is its relative low cost. It is not commonly found in hi-fi component systems.

The *belt drive* mechanism conveys the shaft motion of the motor to the turntable via a belt of rubber, plastic, or polyurethane. The belt effectively decouples motor vibrations from the turntable and thereby reduces rumble. Most commercially available turntables are belt driven.

Most recently, a *direct drive* mechanism has appeared on the market. The axis of the platter is connected directly to the shaft of the motor. The electric motor is controlled by an electronic mechanism that keeps the rotational speed of the shaft constant even in the face of slight voltage fluctuations occurring in the 110 volt line source. The direct drive turntables are more expensive, but their specifications are somewhat better.

A second requirement of obvious importance: once the platter is rotating steadily and quietly, the tip of the stylus must be brought into contact with the record groove. This is the function of the tone arm. The tone arm is *not* "just another part" of a turntable. It is a precisely engineered device designed and manufactured as a separate component in its own right. Fine, precision bearings are used in the pivot and suspension assembly of the tone arm so that it can swing freely across the record with minimum friction.

For perfect reproduction the center line of the cartridge should be

*The relative merits of turntable vs. changer are discussed in Chapter 16.

exactly parallel to the tangent line of the record groove. This configuration permits the stylus to move from side to side along a line exactly 90° to the tangent line of the record groove. These are the ideal circumstances. Departure from this ideal is referred to as *tracking error*. Because the tone arm is pivoted at the rear, the cartridge moves in an arc across the record. At only one point in the movement of the stylus is its tracking error zero. This point is usually chosen to be on the center groove of a record album. At both larger and smaller radii, there is an error. There is a tendency to lengthen the tone arm since a long arm flattens the travel arc and minimizes tracking error.

Ideal stylus motion
Tangent line
Tracking error angle

A tone arm is delicately balanced. The best tone arms will track flawlessly at a tracking force as low as one gram. With low tracking force, however, comes another problem. The tone arm tends to skate toward the center of the rotating platter. At very low tracking forces, the tone arm may actually move across the grooves toward the center. *Antiskating* devices counteract this tendency. A slight outward force is exerted on the tone arm by a tiny spring. This antiskating force can be adjusted to meet various cartridge requirements.

Specifications relevant to turntables are defined below. (As we shall see later, the specifications of a cartridge must be considered with respect to the specifications of a turntable.)

1. Wow and Flutter (in Percent)

Wow is a slow fluctuation and flutter is a high-speed quiver. Both are consequences of variations in the rotational speed caused by deficiencies in the drive system of the turntable. Both are expressed as a percent variation from constant speed. The smaller the percent figure, the better. For hi-fi performance the wow/flutter percentage cannot exceed 0.15 percent.

2. Rumble (in Decibels)

Rumble is the amount of noise generated by the motor and turntable assembly that is transmitted to the cartridge and amplified by the electronics of the system. The specification given as −60 db means that

the rumble is 60 db below the desired signal. A rumble figure of −45 db or smaller is unacceptable for hi-fi systems. (Note that rumble is most pronounced at low frequencies. In other words, if the speakers of a sound system have a good low-frequency response, rumble can be a troublesome factor.)

3. Platter Weight (in Pounds)

The larger the platter weight, the greater is the rotational inertia of the platter and the more constant is the rotational speed. Weights range from about 2 to just under 10 lbs.

4. Stylus Tracking Force Range (in Grams)

The counter balance on the tone arm permits the tracking force to be varied over a range. This range typically extends from 0 to 5 grams.

5. Maximum Tracking Error (in Degrees/In.)

The tracking error can be set to zero for a particular groove on the record. Commonly, the zero setting is made for a groove whose radius is intermediate between the radii of the outermost and innermost grooves. For example, on a $33\frac{1}{3}$ rpm record album the outer groove has a radius of approximately 14.5 cm (5.7 in.) and the inner groove a radius of 6.1 cm (2.4 in). A groove midway between these extremes has a radius of about 10.3 cm (4 in.).

Suppose the tracking error is zero when the stylus is tracking on a groove of radius 4 in. For grooves with both larger and smaller radii there will be a tracking error. The tracking error for commercial turntables range from 0 to 2 degrees/in. If the specification reads 1 degree/in., we can conclude that the tracking error will be 1° for a 5-in. groove or for a 3-in. groove. What would the error be for a $5\frac{1}{2}$ in. groove?

Obviously, the best turntables limit the tracking error to the smallest possible value.

The Tape Deck Information can be stored on a record disc, it can also be stored on magnetic tape. The storage of information on magnetic tape is based on the same electromagnetic principles discussed earlier. As was indicated then, a magnetic field is created by an electric current. Because of the electron configuration in a few select atoms, such an atom can give rise to its own magnetic field. An example of such an atom is iron. In an ordinary piece of iron these atoms are randomly oriented and the iron piece displays no unusual properties. If, however, these iron atoms can be oriented so that they are aligned, the iron will exhibit the properties of a magnet.

Recording tape consists of a thin cellulose acetate base coated with a much thinner layer of iron oxide. This coating can store information. Here is how it works.

An audio signal is transformed into an electrical signal by a microphone transducer. The conductor carrying the electrical signal is coiled around an iron ring in which a small gap has been cut. (See the figure below.) The magnetic field resulting from the electrical current is largely contained within the volume of the iron core itself and within the small gap. However, and here is the crucial point, the magnetic field in the gap alternates in accordance with the rhythm and intensity of the incident audio signal. In other words, the magnetic field "follows" the audio signal. When magnetic tape is passed through (or by) the gap in the ring, the alternating magnetic field prods the iron atoms into an alternating alignment thereby "freezing" information on the tape.

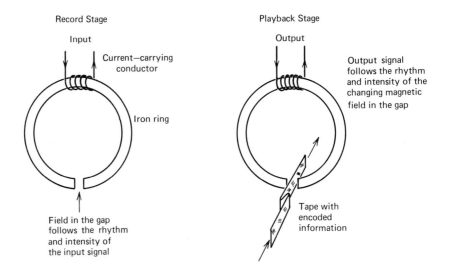

Tape on which information has been encoded has a magnetic field of its own. The character of this magnetic field is determined by the way in which the atoms are aligned in the recording process. To retrieve this information in the playback process, the tape is passed through (or by) a similar gaplike structure—called the playback head. This time, the magnetic field of the tape sets up an alternating magnetic field in the iron core and this field induces an electric current in a conductor wound around the core. The electrical signal, containing the same audio information as the original recorded signal, can be amplified and transformed by a transducer into an acoustical signal.

In early recording machines, the full width of the tape was used for one input signal. Later, the tape was split into two and still later into four "tracks." With stereo tapes, two separate signals are recorded by means of two recording heads. If two-track tape is used, the tape must be rewound to be played. Four-track tape eliminates the rewind step as two

Two—track tape Four—track tape

sets of tracks are recorded in opposite directions. Playback also involves two separate magnetic heads.

Once again we can ask, what's important?

Tape quality is as important as the equipment itself. Furthermore, tape should not only be matched to the equipment with which it is to be used, it should also be matched to the program material to be recorded. Quality tape possesses the obvious attributes of strength, resistance to stretching, maintaining flexibility, and smoothness. (Rough tape can shorten the life of a playback head.) Acoustical properties of tape such as its frequency response, dynamic range, distortion, and noise characteristics are determined by the magnetic film. Again, a tape's acoustical properties should be matched to the intended audio uses.

Once a tape has been put into place the tape equipment takes over. The drive mechanism must move the tape past the recording head or playback head at a very constant speed. Speed inconsistencies lead to wow and flutter. Most tape units can vary the speed with which the tape is moved. Speeds of $7\frac{1}{2}$ ips (inch per second), $3\frac{3}{4}$ ips, and $1\frac{7}{8}$ ips are common. For good high-frequency response, fast tape speeds must be used. (Professional studio machines transport the tape at 15 ips.) High tape speeds, however, imply short recording time. Compromises are necessary.

The drive mechanism must also permit rapid forward and backward winding. Some units accomplish these requirements with one motor. Higher precision is obtained with two and three motor systems where the supply reel, the take-up reel, and the capstan are each driven by a separate motor. (The rotating capstan moves the tape at constant speed.)

Tape machines have at least two different heads: a record/play head and an erase head. Better machines separate the record/play roles and employ individual heads for these functions. The reason for this is that different head gaps are ideally required for the recording and playback. A head that combines both functions necessitates compromise. Anyone planning to engage in serious recording should have a three-head system. (A three-head system permits the monitoring of the playback while it is being recorded.)

If unrecorded tape is passed by the playback head, a pronounced hissing sound is heard. Even when the tape has a recording impregnated on it, the hiss is still present. The hissing sound is masked by musical passages at the fortissimo level; however, at pianissimo levels the hiss is

aggravatingly distracting. Until recently, this inherent noise has been a serious drawback to the taped program source.

The Dolby noise reduction system electronically eliminates the tape noise, circuit noise, etc. without sacrificing the original tonal quality. The Dolby system must be employed *both* in the recording phase and the playback phase. In the recording step the high-intensity signals are left alone. The low-intensity signals are electronically upleveled as illustrated below. Compensation must occur in the playback step; namely, the low-intensity signals must be downleveled so that the same relative intensities exist in the playback as existed in the original musical rendition. This downleveling of the low-intensity signals is done electronically but in the process the noise is downleveled an equal amount, as shown in the illustration. This means that the signal-to-noise ratio is large even for low-intensity signals.

Record step

Playback step

The specifications defined below are commonly cited for tape equipment.

1. Tape Speeds (in Inches per Second)

This specification is given for open-reel tape recorders.

2. Number of Heads and Numbers of Motors

Again, this is given for open-reel equipment.

3. Frequency Response and Uniformity (in Hertz and Decibels)

See pages 222 and 223.

The frequency response of tape equipment is dependent on the tape speed. At speeds of $7\frac{1}{2}$ ips, the frequency response should be a least 30–16,000 Hz \pm 2 db for hi-fi work.

4. Wow and Flutter (in Percent)

See page 226.

Quality open-reel recorders have values as low as 0.04 percent while cassette recorders are in the range 0.1–0.3 percent.

5. Signal-to-Noise (in Decibels)

The signal-to-noise ratio is a measure of a device's freedom from self-created and other unwanted noise. The ratio is expressed in decibels and the higher the value the better. Open-reel and cassette systems have values in the range of 55 to 70 db and 40 to 60 db, respectively. Cassette recorders without a Dolby noise reduction system are unacceptable as a hi-fi component.

6. Channel Separation or Crosstalk (in Decibels)

See page 223.

Channel separation should be 55 db or greater.

The AM/FM Tuner

Of the program sources we shall consider, the tuner is unique. It is entirely electronic with no motors driving wheels, gears, or shafts. The tuner is available as an independent component or in combination with a preamplifier and amplifier—the composite whole is called a receiver.

The information in record albums is contained in the wiggles of the record groove while for tape it is contained in the wiggles of the magnetic orientation at the atomic level. The information utilized by tuners is contained in the wiggles of electric and magnetic fields picked up by a receiving antenna. Radio waves, light, X rays, ultraviolet radiation are all electromagnetic waves differing only in their frequencies and wavelengths. All propagate with the incredible speed of 3×10^8 m/sec. Radio stations transmit over two frequency ranges:

AM Band	53,500	to	1,605,000	Hz
	53.5	to	1,605 kilohertz	kHz
FM Band	88,000,000	to	108,000,000	Hz
	88	to	108 megahertz	MHz

The wavelengths associated with these bands can be calculated from the relationship $v = 3 \times 10^8$ m/sec $= f\lambda$:

AM Band	561	to	187	meters
FM Band	3.41	to	2.78	meters

Information is conveyed in radio waves by two techniques: amplitude modulation (AM) and frequency modulation (FM). Each technique involves a *carrier frequency*—a different frequency for each transmitting station in a particular geographic area—that is ultimately "tuned in" by the home audience. Many frequencies are simultaneously incident upon the receiving antenna. To tune in a particular frequency, the familiar concept of resonance is utilized. Electrical circuits can be designed to resonate electrically. The resonate frequency of the circuit can be varied by changing the value of tunable components in the circuit.

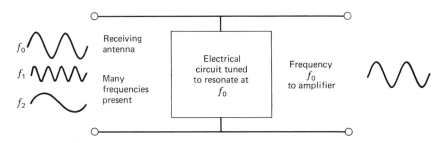

In radio transmission, audio information is superimposed on the electromagnetic carrier wave. In AM, the audio information alters the amplitude of the carrier wave and leaves the frequency alone. The process is illustrated in a schematic way below. In FM the audio information alters the frequency and leaves the amplitude alone.

FM offers distinct advantages over AM. First, the Federal Communications Commission (FCC), the regulatory agency for broadcasting, imposes an upper limit to the frequency that an AM station can transmit. This limit is 5000 Hz, which is hardly adequate for sound reproduction with fidelity. FM, on the other hand, transmits audio information with frequencies ranging from 50 Hz to 15,000 Hz. Second, the dynamic range in AM has

Amplitude modulation

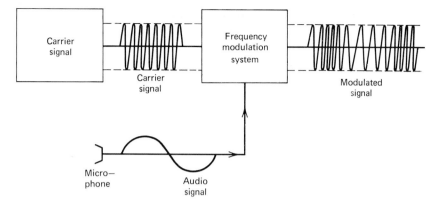

Frequency modulation

definite limitations imposed by physical principles whereas in FM the dynamic range is potentially so broad that limitations must be imposed by the FCC.

In spite of the advantages of FM over AM, it was not until stereo-FM was developed that the FM tuner became a valuable program source for the hi-fi enthusiast. In stereo broadcasting, two independent sets of information are transmitted. The process by which this is accomplished is called *multiplexing*. A station broadcasting stereo FM transmits three signals: a main carrier frequency, a subcarrier frequency removed from the main carrier by 38 kHz, and a pilot frequency removed from the main carrier by 19 kHz. Let us symbolize one set of information by L (for left) and the other information set by R (for right). The main carrier contains the sum of both information sets, that is, $L + R$. The subcarrier frequency contains the difference between the two information sets: $L - R$. A tuner that is only equipped to receive monaurally will utilize the $L + R$ signal. In stereo reception, the two signals are electronically *added* and *subtracted*, thereby retrieving each separate set of information:

$$(L + R) + (L - R) = 2L \rightarrow \text{to left channel}$$
$$(L + R) - (L - R) = 2R \rightarrow \text{to right channel}$$

The pilot frequency activates the multiplexing circuits in the tuner and provides a frequency and phase match between receiver and transmitter.

The performance specifications of a tuner can be anticipated if the demands made upon a hi-fi tuner are identified. Certainly the frequency response must embrace the range of frequencies transmitted and do so uniformly. If the tuner is to be used in a location far removed from transmitting antennas, it must be capable of detecting and receiving a very weak signal. On the other hand, if the tuner is to be used in an urban area with many competing signals, it must be capable of extracting the desired signal while suppressing adjacent signals and noise. Furthermore,

suppose the tuner is to be used in a location between two FM stations transmitting at the same frequency. It would be desirable if the tuner could identify the stronger of the two signals and work with it while at the same time suppressing the weaker signal.

Once signals have been received they become the input ingredient for electronic circuits. The circuitry must maintain the integrity of these signals. The left and right channels must be kept separate, any distortion of the signal must be kept to a minimum, and electronic noise must be minimized so that the signal-to-noise ratio is large. The tuner specifications defined below reflect these various demands.

1. Frequency Response

Not so crucial here as long as it covers the 50 to 15,000 Hz range transmitted by FM stations. Typical tuner responses are 20 to 15,000 Hz ± 1 db.

2. Sensitivity (in Microvolts, μV)

The sensitivity indicates the minimum antenna signal strength that can be successfully converted into a sound signal by the tuner. Sensitivity ratings of component-type tuners range from 1.5 μV (outstanding and costly) to 4.0 μV.

3. Capture Ratio (in Decibels)

If two signals of the same frequency are incident upon the receiving antenna, the capture ratio expresses the ability of the tuner to suppress the weaker of the two signals and enhance the stronger. Capture ratios range from about 1.0 to 5.0 db. The smaller the capture ratio, the better.

4. Selectivity (in Decibels)

This measures the tuner's ability to select one frequency from nearby frequencies. This specification ranges from 45 to more than 100 db. This number should be large.

5. Stereo Separation (in Decibels)

For a stereo effect the two signals (L and R) must be kept separate—particularly in the midrange domain (400 to 1000 Hz). This specification is usually cited for a particular frequency; for example, separation at 1000 Hz typically ranges from 35 to 50 db. The larger the number, the better.

6. Total Harmonic Distortion, THD (in Percent)

Distortion, in its broadest sense, is anything that adulterates the program signal. Electronic circuits have the capacity for producing harmonics of a signal, that is, if the signal has a frequency f, the circuits can produce frequencies $2f, 3f, \ldots$. Obviously, any signals in the output not present in the input are undesirable. The THD measures in percentage the unwanted signal vs. the wanted signal. The lower the percentage the better. For tuners the THD ranges from 0.1 to 2 percent.

Input: *f* | Component | Output: *f*

No harmonic distortion

Input: *f* | Component | Output: *f, 2f, 3f, . . .*

Harmonic distortion present

7. Signal-to-Noise, S/N (in Decibels)

This is a ratio, in db, comparing the wanted signal and the unwanted noise. A S/N of 60 db means that the signal is 1,000,000 times stronger than the noise. Obviously, this specification should be large. The S/N of tuners range from 50 to 80 db.

HIGH-FIDELITY PROGRAM CONTROL AND AMPLIFICATION

Each program source delivers an electrical signal containing information. The ultimate objective is to transform this electrical signal into an acoustical signal. However, the transducers that perform the electrical-signal-to-acoustical-signal transformation require hefty electrical signals. The signals produced by the program sources are just too weak and therefore must be enhanced. This enhancement is the function of the control and amplification components of a sound system.

The two components making up this part of a sound system are the preamplifier and the power amplifier. There are various ways manufacturers package these two components. They are available as separate units. They can be combined into a single unit, which is called an integrated amplifier or control amplifier. Or a tuner can be combined with these two units and the combination is called a receiver. Outstanding high-fidelity performance can be obtained with any of the forementioned combinations; however, there are some pros and cons concerning these options that we shall discuss in Chapter 16.

The Preamplifier

The preamplifier is the hub of the sound system. The signals from all the program sources are directed via cable connections into the preamp; from the preamp go the signals that ultimately drive the loudspeakers. The preamplifier is the control unit of a sound system.

The first task of the preamplifier is to amplify the input signals. The three principal inputs to the preamp are labeled phono, tape, and tuner (there may be several tape inputs for cassette-type and cartridge-type equipment). The phono input is a nonlinear input, which distinguishes it from the others. Earlier it was pointed out that in the record-cutting process compensatory steps are taken at both low and high frequencies. The response of the preamp to the phono signal is designed to equalize the signal so that all frequencies have the same relative intensities as they had prior to the compensatory steps taken in the recording process. The signals coming into the preamp are very weak. These signals must be selectively amplified while unwanted noise and hum are kept small.

The second task of the preamp is to serve as the control center of the sound system. There are many knobs and buttons that can be manipulated to influence the resulting sound. Some of the controls and their functions are outlined below.

1. **Function Selector.** The function selector chooses between the available program sources; for example, phono, tape, AM, or FM.

2. **Speakers Switch.** Many preamplifiers have provisions for connecting two or more *pairs* of speakers, which can then be driven separately or in pairs. Such a facility permits speakers to be placed in a location remote from the sound system itself. Also, different speaker systems can be compared by switching back and forth from pair *A* to pair *B*.

3. **Volume Control.** This control is self-explanatory.

4. **Balance Control.** The relative volume in the two channels of a stereo system can be regulated by this control. This adjustment permits a sound system to be better matched to a particular room environment.

5. **Tone controls.** Tone controls can shape the perceived sound either by depressing the treble frequencies relative to the bass frequencies thereby giving the illusion of "boosting" the bass or by depressing the bass frequencies relative to the treble frequencies. (See the illustration below.)

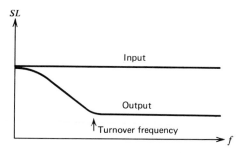

Bass "boost." Actually, the treble frequencies are attenuated.

Tone controls are frequently abused—particularly by the hi-fi novice. Heaving, boomy bass accentuation is the most common abuse. The purpose of the tone control is to compensate deficiencies in speaker response, deficiencies in program source, and peculiarities of the room environment. We learned, for example, that the surfaces of upholstered furniture, carpets, and draperies absorb high frequencies more efficiently than low frequencies. In such an environment it might be desirable to "boost" the treble frequencies

Some preamplifiers have tone controls that simultaneously modify both channels. More expensive units have separate controls for each channel. Since the tone control circuits can introduce a degree of distortion to the signal, it is best to use controls sparingly and leave them near the neutral level.

6. **Loudness Control**. This is not to be confused with the volume control. At low-intensity levels, the human hearing sensitivity drops off at both high and low frequencies—see the loudness-level curves on page 42. The loudness control compensates for the drop-off of hearing sensitivity by applying high- and low-frequency boosts to the audio signal. This control should not be activated when listening at high volume levels.

7. **Filter Controls.** High- and low-frequency filters cut out the sound spectrum above and below certain frequencies. The high-frequency filter is used to eliminate scratching and hissing noise from worn records and tapes (a Dolby noise reduction system does this for tapes). The low-frequency filter cuts out hum and rumble generated by turntables. Since these filters also eliminate program material, they should only be used when necessary.

The significant specifications of a preamplifier can be understood in terms of the tasks it must perform.

1. **Frequency Response (See page 222 to 223.)**
2. **THD. (See page 234.)**
3. **Intermodulation Distortion, IMD (in Percent)**

When signals of frequencies f_1 and f_2 are introduced to a circuit, sum $(f_1 + f_2)$ and difference $(f_1 - f_2)$ frequencies can be produced as a consequence of nonlinear behavior of the circuit. These sum and difference frequencies can be particularly annoying as they are harmonically unrelated to the original signals. This specification is given as the percentage of unwanted signal; therefore, the lower the percentage the better.

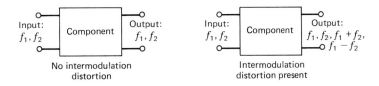

4. **Signal-to-Noise. (See page 235.)**

Frequently S/N specifications are given for different operational modes such as phono, tape, and auxiliary.

The Power Amplifier

To drive the cone of a loudspeaker, audio power ranging from a few watts to a hundred or more watts is required. Since the signals coming from the preamp are small, the amplitudes of these signals must be significantly increased without any other alteration in the waveform. The power amplifier is the component that boosts the signals from the preamp to the level adequate for the loudspeakers' purposes.

A power amplifier delivers its power into a "load." In order for power to be effectively transferred from the amplifier to the load, the output of the amplifier must be matched to the input of the load. The parameter that is cited as an indicator of the output and input character of a component is the *impedance*. Most speakers—the load of a sound system power amplifier—have impedances of 8 ohms. (Some speakers have a 4-ohm impedance.) Therefore, the appropriate output impedance would be 8 ohms.

Impedance matching between output and input is necessary.

There is perhaps no other component over which more confusion exists and over which more commercialized distortion occurs than with the power amplifier. The problem is that there are several ways—most of them misleading—that power ratings can be specified. Suppose the input to an amplifier is a pure tone. The output will also be a pure tone with a larger amplitude. This is illustrated below. Notice that the output signal starts with a large positive value (point *a*), goes through zero (point *b*) to a maximum negative value (point *c*), and back through zero to a maximum positive value (point *e*). The question now is: how should the output be rated? In terms of the value at *a* or *e*? It can be so rated; however, such a *peak rating* is misleading. It makes more sense to define a rating that reflects the "effective" output rather than the peak output. Such a rating is the root-mean-square (RMS) output, which represents the effectiveness of an alternating output signal in terms of its peak value.

A power amplifier should be capable of delivering its power *uniformly*

over a specified frequency range. Sometimes power ratings are given at a particular frequency—for example, 50 watts at 1000 Hz—with no indication of its power capabilities at other frequencies. The power rating should be coupled with a distortion figure. Power can be produced by an amplifier, but in the process the input waveform can be butchered. For high-fidelity applications, the integrity of the waveform must be maintained. Thus, distortion must be kept to an acceptable level. The power rating should be specified in terms of the load impedance. If 60-watt power is desired, the power amplifier must be capable of delivering those 60 watts to the intended load. Finally, the amplifier should be capable of sustaining its rated power output on a continuous basis. One is *not* interested in the "instantaneous peak power" (IPP rating), but the continuous power capabilities of an amplifier.

Because the potential for confusion (and chicanery) are so great, the FTC (Federal Trade Commission) has legislated standard formats that manufacturers are supposed to follow. This legislation was put into effect November 4, 1974. In this federal ruling, not only is a format recommended, but also the conditions under which the amplifier is to be tested and its specifications determined.

There is controversy over how much power is desirable. The consumer is presented with the choice of power outputs ranging from 10 watts to "superpower" amplifiers in the 200 to 1000 watt domain. In Chapter 16 we shall discuss the question of how much power is needed, but for now some of the hazards of superpower should be understood.

Cost is a first consideration. As a rule of thumb, the cost of a unit increases with its power capability. Speakers can be damaged when too much power is applied. Furthermore, the damage can occur at rather normal listening levels. For example, if the tone arm is accidently dropped, or if the preamp selector switch is changed (going from phono to tape), a sudden burst of power resulting in a loud "pop" can result. Such a "pop" can blow a speaker.

The specifications of power amplifiers are outlined below.

1. Power Output (in Watts).

This specification should appear in the following form:

> 60 watts RMS continuous power per channel into 8 ohms, 20–20,000 Hz ± 1 db, with THD < 0.5%.

Stated this way, the power specification includes the *frequency response* and the THD. Note the words RMS, continuous, and per channel (sometimes specifications give a total power figure that is the sum of both channels).

2. Music Power or Dynamic Power

A rather bogus specification given in place of continuous power output.

3. Power Bandwidth

This specification complements the power and frequency response specifications. The power bandwidth is the frequency range over which the power amplifier will deliver half its rated output power without exceeding its THD specification.

4. Signal-to-Noise (in Decibels)

See page 235.

5. Intermodulation Distortion (in Percent)

See page 237.

6. Damping Factor

This is a measure of the unit's ability to control speaker motions caused by forces other than the audio signal. High values are better.

HIGH-FIDELITY PLAYBACK

The character of the overall sound coming from a component system is determined more by the speakers than by any other part of the system. It is not surprising, then, that the topic of speaker systems evokes more controversy among audiophiles than any other subject. The components we have discussed so far can be evaluated primarily on the basis of their specifications. This is not so much the case with speaker systems: hence, the controversy.

A loudspeaker is required to transform electrical power into acoustical power. It should do this efficiently and uniformly over a wide frequency range, it should disperse this sound energy uniformly over the listening space, and it should respond faithfully to sudden changes in the electrical signal. These are obvious operational requirements that form the basis for making an objective performance assessment. In the final analysis, however, subjective elements enter the consideration, which makes the choice of a speaker system a personal matter.

A loudspeaker is a transducer. It operates in a manner converse to the operation of the cartridge; namely, it transforms electrical signals into acoustic (i.e. mechanical) signals. This transformation can be achieved in several ways. Most commonly, an interaction between an electrical current and a magnetic field is employed. Such is the case with the *dynamic loudspeaker*.

Consider the following figure. In parts (*a*) and (*b*) cross-sectional representations of the heart of the dynamic loudspeaker are shown. A coil of wire is slipped over and coaxial with the north pole of a yoke-shaped magnet. When current passes through the coil in one direction, a force is exerted on the coil tending to pull the coil out of the yoke, see (*a*). A current in the other direction gives rise to a force pushing the coil into the yoke; see (*b*). Thus, an alternating current successively pushes and pulls the coil into and out of the magnetic yoke.

In the dynamic speaker, the coil is connected to a cone, as illustrated in (*c*). As the coil responds to the forces acting upon it, the cone moves in and out, thereby disturbing the air and initiating a sound wave.

Dynamic speakers come in different sizes. The large speakers—10 to 12 inches in diameter—are called *woofers*. The smallest ones, known as *tweeters*, are approximately three inches in diameter. The *mid-range* speaker is intermediate in size—in the 4 to 5-inch diameter range. This difference in size is necessary for two reasons. First, to produce a certain amount of acoustic power, say 1 watt, a certain amount of air must be moved. Specifically, more air must be moved at low frequencies. This can be accomplished either by increasing the amplitude of the cone's motion or by increasing its size. The latter option is chosen. Woofers are best suited for low-frequency (up to 500 Hz) signals, tweeters for high frequencies (3000–20,000 Hz), and midrange speakers for the intermediate frequencies.

The second reason for different-size speakers is due to the way the cone moves. It is desirable to have the cone move as a unit. At low

(*a*) (*b*)

Suspension

Cone

⊗ Current into paper
⊙ Current out of paper
F Force on current coil
B Magnetic field

(*c*)

frequencies, the cone material is sufficiently stiff so that it moves as a whole. At high frequencies, however, waves are initiated at the center of the cone and move in wave fashion toward the rim. The result of these waves is to produce irregularities in the frequency response of the speaker. By having small speakers for high-frequency signals, the problem is reduced.

A second method for transforming an electrical signal into an acoustic signal involves the interaction between oppositely charged objects. This is the mechanism employed by the *electrostatic loudspeaker*. In this speaker system, the charge on two parallel plates varies according to the input signal. As the magnitude of the charge varies, one of the plates moves in and out and disturbs the contiguous air.

The dynamic and electrostatic loudspeakers radiate sound directly into the air. There are disadvantages to this direct-radiation process. Among them are the low efficiency of the process and the narrow directivity pattern of the resulting sound field that occurs at high frequencies. These disadvantages can be circumvented by the introduction of a horn between the coil driving unit and the air. The horn couples the loudspeaker and the air making the transfer of acoustic energy to the air a more efficient process. Furthermore, the sound is dispersed much more uniformly from the horn. This is particularly true at high frequencies making horn-type coupling quite common for tweeters.

Horn-type coupling of the type employed with tweeters

An isolated loudspeaker vibrating alone produces very little sound. The reason for this is that the waves formed in the front and back of the loudspeaker can effectively cancel each other out. When the speaker cone is thrust forward as in (a) in the figure below, a high pressure compression is formed in the front and a low pressure rarefaction is formed in the back of the cone. If the wavelength of the sound is large compared to the dimensions of the speaker, an air flow will be set up between the high pressure and low pressure regions with the result that the sound intensity is substantially reduced.

This difficulty can be solved by mounting the loudspeaker in a baffle—for example, mounting it in a wall. The baffle prevents the air in front from communicating with the air in back of the speaker. A baffle is

(a) (b)

Path length, l

Cancellation does not occur if $l > \lambda$

Baffle

effective as long as the resulting path length between the front and back of the speaker is greater than the wavelength of the sound. In other words, the time required for a disturbance to travel from the front to the back must be greater than one period of the speaker-cone's motion.

Speakers are not normally mounted in baffles. Rather, speakers are mounted in an enclosure. Such an arrangement prevents the transport of air from the front to the back of the speaker, but other problems arise. One of the problems is due to the resonant properties of the air in the enclosure. In Chapter 12 we saw that the air in a box of dimensions l_x, l_y, l_z has resonant frequencies given by

$$ f = \frac{v}{2} \sqrt{\left(\frac{n_x}{l_x}\right)^2 + \left(\frac{n_y}{l_y}\right)^2 + \left(\frac{n_z}{l_z}\right)^2} \qquad n_x, n_y, n_z = 0, 1, 2, \ldots $$

Air in the enclosure vibrating at large amplitudes at these resonant frequencies can not only damage the cone of the loudspeaker, it can also excite the sides of the enclosure into vibration. The vibrating sides of the box initiate large amplitude sound waves outside the box, thereby distorting the perceived sound.

The *air-suspension* enclosure is a sealed cabinet. (A slow air leak must be provided so that the speaker cone can maintain its neutral position as the atmospheric pressure changes.) The enclosure is constructed out of rigid materials and it is braced to prevent the sides from vibrating. Furthermore, the enclosure is stuffed with a material such as glass wool or plastic foam to absorb the sound radiated from the back of the speaker. The woofer in the air suspension speaker system is mounted with a soft suspension. The air in the enclosure provides the "spring," thereby replacing the mechanical spring of the cone suspension. This arrangement permits the cone to vibrate with larger in–out amplitude and improves the base response. Because of the woofer suspension and the internal stuffing, air-suspension speaker systems are relatively inefficient; therefore, at least 20 watts of amplifier power are required.

The optimum cabinet size for the air-suspension system encloses about two cubic feet of air. This size is conveniently small thereby giving rise to the generic term "bookshelf" speaker.

The *bass-reflex* speaker system makes use of the sound generated at the back of the loudspeaker. In this speaker system, an opening called the *port* is made in the front of the enclosed box. The air inside the enclosure is excited by the speaker-cone motion in much the same way as the air in a bottle is excited when one gently blows over the opening. When the system is properly designed, the port becomes a second "cone" driven by the back side of the loudspeaker.

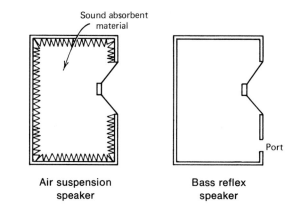

Sound absorbent material

Air suspension speaker

Bass reflex speaker

Port

To be effective, the sound emanating from the port must reinforce the sound coming from the loudspeaker, not detract from it. Remember that a compression is formed in front of the loudspeaker at the same instant a rarefaction is formed at the rear of the loudspeaker. In other words, the two waves initiated are 180° out-of-phase. For reinforcement, a crest emanating from the port must coincide with a crest emanating from the speaker. Such will be the case if the "effective" path length between the back of the loudspeaker and the port is $\lambda/2$. Then a crest leaving the loudspeaker will coincide with the crest of the previous wave as it leaves the port.

Bass-reflex speaker systems have the reputation of suffering from a boomy, artificial bass. Although this reputation has been earned, it is undeserved. The air resonance of an improperly designed system can be so pronounced that the bass response of the system is very uneven. However, when the loudspeaker is properly adapted to the enclosure, when the proper amount of damping is added, the bass-reflex system can produce a smooth wide-range bass response.

The port is only functional at low frequencies. At higher frequencies, the wave arrives at the port (from the rear of the speaker) out-of-phase with the wave leaving the front of the speaker. To prevent this

undesirable situation, the enclosure is packed with a material that is very absorbent for high frequencies but passes low frequencies.

As we have seen, speaker systems can have three or more loudspeakers per cabinet. A *crossover network* electronically directs the right frequencies to the right loudspeakers. The signal coming into the speaker contains all frequencies. A crossover network can extract all frequencies up to 600 Hz and direct them to the woofer. A second crossover network directs frequencies from 600 to 5000 Hz to a midrange speaker. The remaining frequencies go to the tweeter.

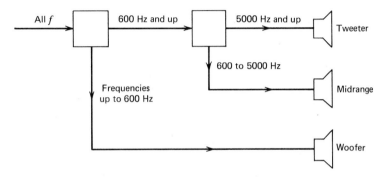

Crossover network

Specifications for speaker systems are somewhat less significant from the consumer perspective than is true for other components. Decisions pertaining to speakers are made on a more subjective basis. The specifications identified below, however, are quite useful.

1. Frequency Response and Uniformity of Response. (See page 222.)

Note. This specification is not as good for loudspeakers as it is for other components. The uniformity of response tends to fall off below 100 Hz and above 8000 Hz. Many manufacturers do not include uniformity of response in their specifications. A typical specification might read 35–20,000 Hz ± 5 db.

2. Power Handling Capacity (in Watts)

The power that the speaker system can handle on a continuous basis without damage is indicated by this specification.

3. Impedance (in Ohms)

Although this gives us no information about the sound of a speaker, it is important. The output impedance of the power amplifier should be matched to the input impedance of the speaker if efficient transfer of energy is to occur. Impedances range from 4 to 16 ohms for commercial speaker systems.

(If two sets of speakers are used simultaneously, it is important to

remember that impedances in parallel reduce the impedance. For example, a parallel combination of 16-ohm speakers reduces the total load impedance to 8 ohms.)

For Further Thought

1. Illustrated below are two configurations of a stylus (*S*) to which is attached a magnet (*M*). Next to the magnet are two coils of wire (*C*). In one case (*a*) the stylus is stationary. In the other case (*b*) the stylus is moving. In which case is an electrical current induced in the coils? Why?

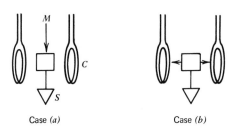

Case *(a)* Case *(b)*

2. Explain why a stylus must move faster in tracing out curve *B* than it does in tracing out curve *A*. (See page 220.)

3. Distinguish between *turnover* frequency and *transition* frequency.

4. For a particular cartridge the desired voltage is 10 times greater than the undesired voltage when measured at 1000 Hz. What is the channel separation for this cartridge?

5. A stereo record has information stored upon it. How is that information retrieved by a stereo sound system?

6. How is information stored on magnetic tape? How is the information retrieved?

7. A tape deck carries a signal-to-noise specification of 70 db. The signal intensity is greater than the noise intensity by what factor?

8. A radio station broadcasts at a frequency of 1120 kHz.
 (a) Is it an AM or FM station?
 (b) What is the wavelength of its signal?

9. How is information conveyed by an AM broadcasting station?

10. How is information conveyed by a FM broadcasting station?

11. What is the distinction between the volume control and the loudness control on a preamplifier?

12. Compare the nonlinear behavior of the ear with the nonlinear behavior of a power amplifier.

13. All points on a vibrating speaker cone should move together; that is, all points should move in phase with each other. Why is it that a small diameter speaker cone is better suited for high-frequency signals?

14. A loudspeaker is a transducer. Compare the ways a dynamic loudspeaker and an electrostatic loudspeaker transform a signal of one type into an acoustical signal.

15. A woofer is mounted in a circular baffle having a diameter of 2.5 meters. Suggest the range of frequencies for which this baffle is effective.

15 Sound Systems and Their Environment

Sound is reproduced by the components we discussed in Chapter 14. To achieve a high degree of fidelity and realism, these components must:

(a) Possess a frequency range that includes all the audible components of the sounds being reproduced.

(b) Possess a distortionfree intensity range that embraces the intensity range associated with the recorded sounds.

(c) Preserve the spatial sound pattern of the original sound.

(d) Approximate the reverberation characteristics of the original sound.

Monophonic sound systems are able to meet demands (a) and (b), but do not meet demands (c), and only approximates demand (d). Stereophonic sound systems satisfy demands (a), (b), and (c); demand (d) is only partially satisfied. Quadraphonic sound systems satisfy all requirements, in principle at least.

**SOUND
SYSTEMS**

Three sound system types—monophonic, stereophonic, and quadraphonic—are illustrated in a highly schematic fashion in the following figures. In these figures, the recording and play back details have been completely ignored. Rather, the most crucial differences are emphasized. In the monophonic system, the original sound is picked up by one (or more) microphone(s). Regardless of the number of microphones, however, the significant fact is that only one signal is recorded and only one signal is involved in the playback stage. Monophonic sound systems are employed in disc phonographs, television, motion pictures and radio. Because only one signal is involved, monophonic systems do not satisfy requirements (c) and (d) outlined on the previous page.

A stereophonic sound system employs two or more microphones to pick up the original sound. The spacing between these microphones is

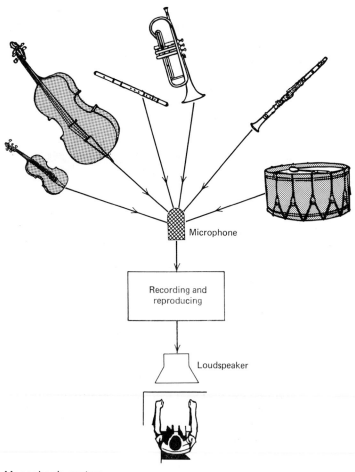

Monophonic system

large enough so that the acoustic signals reaching them are distinctly different. Two different signals are recorded and the individual integrity of these signals is maintained in the playback stage. Stereo sound systems are widely used for home entertainment. Because two signals are employed, stereophonic sound systems provide auditory perspective thereby satisfying requirement (c) above as well as requirements (a) and (b). Requirement (d) is only partially satisfied.

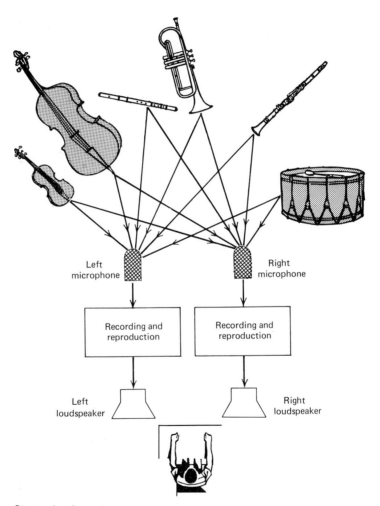

Stereophonic system

Quadraphonic sound systems employs four microphones located in the recording studio so that four distinct signals are recorded. In the playback stage four speakers are situated so that the listener is literally

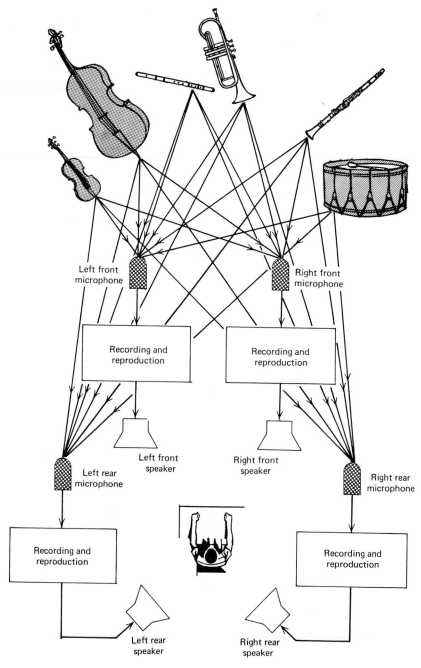

Left front
microphone

Right front
microphone

Recording and
reproduction

Recording and
reproduction

Left front
speaker

Right front
speaker

Left rear
microphone

Right rear
microphone

Recording and
reproduction

Recording and
reproduction

Left rear
speaker

Right rear
speaker

Quadraphonic system

surrounded by sound. All the requirements cited above are in principle satisfied by a quad system. The left and right front speakers provide the spatial perspective while the rear speakers provide the reverberation characteristics. Unfortunately, the present-day quality of quadraphonic recordings do not measure up to the standards of stereophonic recordings.

Sound systems for home use can take many forms. The program sources available are record albums, tapes, and AM/FM signals. These program sources require a turntable, a tape deck, and a tuner for their implementation. A preamplifier (control unit) and a power amplifier prepare the signal for delivery to the loudspeaker. The tuner, preamplifier, and power amplifier can be separate components or they can be combined into a single chassis called a receiver. A monophonic sound system is illustrated in the block diagram below. Here boxes represent components and lines, the interconnections between components. Since there is only one signal involved in a monophonic system, a single cable—two conducting paths—between each component is sufficient.

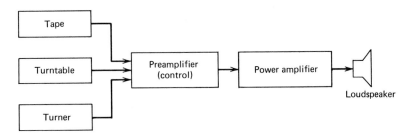

Monophonic sound reproduction system employing separate components.

A diagram of a stereophonic sound system is given below. Although the layouts of the stereo and monosystems are very similar, the fundamental differences are obvious. A stereo system is equipped to

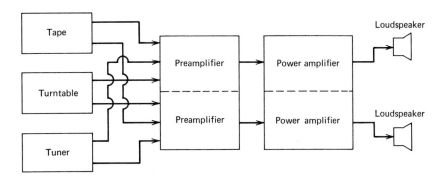

Stereophonic sound reproduction system employing separate components

handle two independent signals. There are in essence two pream-plifiers, two power amplifiers, and there are two loudspeakers. Once again, a stereo receiver combines the tuner, preamplifiers, and power amplifiers into one chassis.

The interconnections are not simply wires; rather, they are shielded audio cables designed to reduce unwanted hum and noise. These cables are usually supplied with components. If not, they can be obtained at an audio dealer. These connecting cables should not be too long. The cable connecting the turntable (or changer) and the pream-plifier, for example, should not be more than six feet in length. On the other hand, the connections between power amplifiers and speakers can be made with ordinary lamp cord and these connecting wires can be long. The lamp cord should be a heavy gauge (18 or 16 gauge) if runs of 15 feet or longer are made.

To protect both the power amplifier and the speakers it is recom-mended that the speaker connections be fused. Fuse holders are available and these can be inserted as illustrated below. For speakers with high efficiency, one-amp fuses are adequate. For very inefficient speakers three-amp fuses will be necessary. For most other speaker systems, two-amp fuses should suffice.

Note. A cable used for connecting components together contains two conducting paths as illustrated above. Into one of these wires, the fuse can be inserted.

A pair of speakers should operate in phase—that is, the speaker cone surfaces should move in the same direction at the same time. (What are the consequences of their moving out of phase?) To test whether they are moving in phase, bring the speakers close together and orient them so that they are facing each other. Play a monaural record with strong bass. After you are very familiar with the musical score and the intensity of the bass, shut the speakers off and reverse the connections on *one* speaker. Replay the familiar musical passage. If the bass seems to be fuller and more intense than it was before, the speakers are now in phase. If the bass seems weaker, the speakers are out of phase and the connections should be returned to their original configuration.

Sound systems that are more elaborate than those illustrated above are clearly possible. For most people, cost determines how elaborate their system can become. An example of a more costly system is given below. This three-way multiamplifier system features a crossover net-

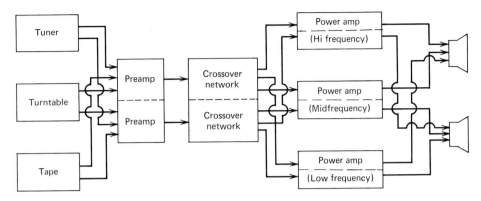

Three-way multiamplifier stereophonic system

work between the preamplifier and power amplifier stages. The signal received from the preamplifier is divided into three frequency ranges: high-frequency range, mid-frequency range, and low-frequency range. Each partial range is then directed to its own power amplifier. From the power amplifiers the signals enter speakers of the proper type (the speaker systems must be equipped with separate input terminals for each frequency range).

COMPONENT PLACEMENT IN THE ROOM ENVIRONMENT

The components of a sound system are placed in a room environment. There are some guidelines that can be cited as to both the relative and absolute placement of these components. The turntable should not be close to the loudspeakers since vibrations from a speaker can be picked up by the tone arm and fed back through the system (a good example of an undesirable feedback loop!). For this reason it would be inadvisable to mount the turntable and the speakers on the same bookshelf or against the same wall. (You can test to determine whether you are getting feedback. Place a record on the turntable and carefully place the tone arm onto a record groove—without the turntable rotating. Now increase the volume control on the preamplifier to a setting typically used for normal listening. If you hear a howl at low or middle frequencies, you probably have an acoustic feedback problem. Try repositioning your turntable and carry out the procedure again.)

Another suggestion concerning relative positioning: if possible, place the control unit (preamplifier or receiver) within arms reach. Having to get up from one's favorite listening chair every time a volume adjustment is made can be vexing.

One "listens" to both the sound system and to the room in which the speakers are placed. Unfortunately, the listening room has fixed features—windows, archways, doors, and dimensions. Few homes are

designed to include a listening room—so we must make do with what we have. Some control, however, is possible. For example, if the ceiling has an acoustical tile treatment, it would be unwise to have carpeting on the floor. Likewise, only one of each pair of walls should be treated with absorbant material such as drapes.

There are no hard-and-fast rules concerning the placement of speakers within a room environment. It is desirable to arrange the layout so that one can sit in front of and between the two speakers. (In a quadraphonic system it is good to have two speakers in front of and two speakers behind the listener.) The two speakers should not be placed too far apart as this can lead to the "hole-in-the-middle" effect. One wants a "wall of sound." Experiment with different arrangements. Remember that the balance control on the preamplifier can be employed to help adapt a system to a particular room configuration.

Most household rooms are rectangular in shape. It is generally best to place the speakers along one of the long walls. They should be located equal distances from both the walls and the corners of the room. If the speakers are facing a draped wall, the resulting sound will tend to be mellower (high-frequency absorption by the drapery material). Removal of carpeting will brighten the resulting sound. Sometimes speakers can be mounted on walls with a resulting improvement in the sound.

Small rooms such as dens or bedrooms pose particular problems. Typically, it is impossible to sit sufficiently far from either of the two speakers. In this situation, one might try positioning the loudspeakers diagonally across from each other in opposite corners of the room.

The best advice that can be given concerning speaker location within a room environment is to experiment. For each speaker configuration, the furniture can be arranged in different ways. One can move around the room to identify whether acoustic dead spots or hot spots exist. The balance control can be varied as well as the tone controls. You, the listener, sit in final judgment on the arrangement that sounds the best.

For Further Thought

1. Construct a block diagram for a monophonic system employing a receiver rather than separate components.

2. Construct a block diagram for a stereophonic system employing a receiver rather than separate components.

3. What are the practical consequences of two speakers in a stereo system moving out of phase with each other? (You might find it useful to review Chapter 6.)

4. What do two stereo speakers moving out of phase with each other have in common with a speaker without a baffle?

5. Represent in diagrammatic fashion the feedback loop between speakers and the tone arm of the turntable.

6. Tests have shown that the sound-intensity level at typical listener locations in several home living rooms varies somewhat with frequency. Specifically, there is a significant valley in the graph of SL vs. f at $f = 200$ Hz. In terms of typical speaker placement within a listening room, can you explain the $f = 200$ Hz dip?

16 Toward a Consummation of the Listening Experience

Careful measurements can be made and specifications can be cited, but in the final analysis sound is reproduced for the human ear and not for the oscilloscope nor for the wave analyzer. Furthermore, if you are buying a sound system, *your* auditory system is the criterion, not that of the shop proprietor.

Recently I witnessed an interesting scene. In an elegant little stereo shop I watched a salesman comparing two power amplifiers for a customer. The salesman was switching back-and-forth between the two amplifiers. The customer kept insisting that he could perceive no difference between the two units in spite of the patient efforts of the salesman who helpfully pointed out the superiority of one of the amplifiers. Soon the atmosphere changed: laughter bordering on derisiveness was directed at the customer as the salesman suggested that the one amplifier was so much better that anyone with ears could perceive the difference. In this case the customer was unusually strongminded and continued to maintain that as far as his ears were concerned, there was no difference. Oh yes, the power amplifier found superior by the salesman cost $1195 while the other unit cost $410.

259

Without doubt, ears differ. A professional musician can hear subleties that go unnoticed by the most of us. The experienced sales man or woman may well develop an unusual keenness of ear. However, since the sound system that graces your residence will be listened to by you, it is your ears that should be the final judge.

Various tests have been carried out to determine listener responses and preferences. For the reproduction of orchestral music, a frequency range of 30 to 15,000 Hz is required for subjectively perfect fidelity. When listeners are given a choice between reproduced sound having a full frequency range (~30 to 15,000 Hz) vs. a restricted frequency range (~100 to 8,000 Hz), there will be almost unanimous preference for the full-range sound.

For subjective perfect fidelity a dynamic range of 70 db is required. The lower limit of hearing is determined by the ambient room noise. A dynamic range of 60 db above the sound level of the ambient room noise is sufficient for high-quality sound reproduction.

The other major criterion for perfect fidelity is the percentage of nonlinear distortion—THD and IMD. The wider the frequency range, the more important it is to keep the nonlinear distortion low. In tests where distortionfree signals were compared with distorted signals it was found that nonlinear distortion of 1 percent could be perceived when the frequency range extended to 15,000 Hz.

Frequency range, dynamic range, and nonlinear distortion are fidelity parameters having psychoacoustic significance. It does not end here, however. The human auditory system is much more subtle. For example, electronic tube systems sound different than solid state (transistor) systems. In describing the difference listeners employ a subjective language: for example, "tube records have more bass," "transistor recordings are very clean but they lack the 'air' of a good tube recording," "with tubes there is a space between the instruments even when they play loud . . . transistors make a lot of buzzing," and "transistors have highs and lows but there is no punch to the sound."

In contrast to these subjective comments, the engineer designing sound reproduction equipment can "prove" there is no difference between tube sound and transistor sound. The proof is based upon impressive specification sheets and electronic facts and figures. Clearly, in terms of the parameters being measured there is no difference in the sound quality. Just as clearly, however, some important parameters are not being measured and these unmeasured parameters are subjectively significant.

In this same vein, speaker evaluation, despite all the many tests and instrument measurements that can be made, is a very subjective activity. A shopper must rely on his or her own hearing to a greater extent for speakers than for any other audio component.

In summary, objective specifications are important and are the place

to start. As one approaches the purchase decision, however, subjective perceptual considerations should play a greater role. Your ears will have to live with the sound reproduced by your sound system.

QUESTIONS FACING THE BUYER

The high-fidelity arena reverberates with the voices of opinionated authorities. Strong arguments pro and con can be expected in response to the questions discussed below; however, there are some principles that can serve to guide the potential buyer.

1. Manual vs. Automatic Turntables?

From a performance viewpoint, wow, flutter, and rumble are important perceptually while tracking force determines record life. There was a time when these crucial specifications were significantly better for the manual turntable. These specifications have now been refined to the point that the distinction between manual and the better automatic units are academic.

Since the manual turntable still has the edge in specifications, the purist will opt for the manual unit. This view is perhaps more emotional than rational. It is true that early automatics compromised specifications for convenience, that their tracking forces were larger, and that the record-changing mechanism was rough on record albums. Automatic turntables now available, however, handle records about as gently as cautious hands.

Economics is a consideration. Manual units offer the most for the money in the budget market. Therefore, a manual should be considered by the bargain hunter assembling a moderately priced system or by the audiophile assembling a "perfectionist" system. For most other people, the convenience of an automatic unit probably outweighs any subtle performance questions.

2. Open-Reel Tape Deck vs. Cassette Recorders?

Once again the question of performance vs. convenience can be raised; however, in this case the distinction is significant. The cassette recorders are convenient; however, performance quality is sacrificed. Fidelity increases with both tape speed and with tape width. Reel-to-reel recorders typically have provisions for two or three tape speeds: $1\frac{7}{8}$ inch per second (ips), $3\frac{3}{4}$ ips, $7\frac{1}{2}$ ips, and 15 ips. The higher the tape speed the better is the frequency response. Cassette units all operate at $1\frac{7}{8}$ ips with the result that those problems alleviated by higher speeds—for example, wow and flutter—must suffer. Furthermore, the cassette tape is only 0.15 inch wide (open-reel tape is 0.25 inch wide) with four narrow recorded tracks. The signal must be compressed on the narrow tape with the attending diminution in the signal-to-noise ratio.

The reel-to-reel format offers greater ease of tape editing and

splicing. Certainly, if serious recording is anticipated, the reel-to-reel recorder is the more desirable unit.

3. Receiver vs. Individual Components?

This question is probably decided on pragmatic grounds rather than performance characteristics. Since both options are available with outstanding specifications, the performances of each type system are essentially equivalent.

The receiver brings together in one cabinet the AM/FM tuner, the preamplifier, and the power amplifier. This consolidation is convenient because less shelf space and fewer connecting cables are required. The total cost of a receiver is lower than the total cost of comparable components purchased separately. However, the initial cost may be higher. The decision to purchase an AM/FM tuner can be postponed, thereby lowering the initial cost if one decides to select individual components.

The system made up of separate components has greater flexibility. As the state of the art changes, one can upgrade the amplifier and leave the tuner alone. With a receiver the whole unit must be traded to achieve a change in any one part. Furthermore, the flexibility inherent in the separate component format permits tailoring a system more specifically to one's needs. A high quality receiver combines a high quality amplifier with a high quality AM/FM tuner. If one resides in an urban area where FM signals are strong, a moderately priced tuner can provide full satisfaction. With separate components, a high quality power amplifier can be combined with a lower quality tuner. In contrast, if one lives in a rural area where tuner sensitivity is a crucial necessity, one can combine a high quality tuner with lower quality amplifier units. (In this way, the total cost of separate components may actually be lower than for the receiver.)

SUGGESTIONS FOR THE PURCHASE PROCEDURE

How does one begin buying? If some guidelines are followed, order can be brought to an otherwise confusing and frustrating experience. Some steps are outlined below.

1. The first step is a budget decision. One should decide how much money can be spent for the system. A limited system of passable quality costs at least $500. From this figure, there is no limit to the amount of money one can spend. Budget restrictions may well determine the number of components in your system—for example, adding a tape deck and an AM/FM tuner—and whether your system is to be stereo or quadraphonic.

 As a buying principle one should strive for a balance of quality. It is a waste of money to couple a superb power amplifier with cheap speakers or to buy a top quality turntable and use a so-so

phono cartridge. To obtain balance, various formulas for dividing the total budget among the various audio components have been proposed. Although there are no hard-and-fast rules, the table below can serve to guide you in beginning the purchase process.

2. The first step must be the choice of loudspeakers. Always begin by auditioning various models within your price range. Take with you to the stereo shop a record album containing the type of music you most enjoy (and an album with which you are familiar) and use your album for auditioning loudspeakers.

Budget Table

System 1:	Stereo phono	Turntable and cartridge	22%
		Preamplifier	22%
		Power amplifier	22%
		Speakers	34%
System 2:	AM/FM stereo phono	Turntable and cartridge	22%
		Receiver	44%
		Speakers	34%
System 3:	Complete stereo system (with receiver)	Turntable and cartridge	14%
		Tape deck (open-reel)	28%
		Receiver	36%
		Speakers	22%
System 4:	Complete stereo system (components)	Turntable and cartridge	14%
		Tape deck (open reel)	28%
		AM/FM tuner	12%
		Preamplifier	10%
		Power amplifier	14%
		Speakers	22%

Compare loudspeakers a pair at a time by having the salesperson switch rapidly back and forth between the two pairs. (This back-and-forth switching process is called A-B testing.) To determine if one pair of speakers is more efficient than the other, have the salesperson adjust the volume control so that you compare the two pairs of speakers under the *same loudness* conditions. (Since louder sound is always more exciting and dramatic, you will tend to judge the loudest speakers as the best speakers.) When you have decided which pair you prefer, compare a different pair with this preference. Continue your pair-by-pair comparison until you have identified the pair that you like the best.

3. With speakers tentatively chosen you know how much of power is required to drive them properly. Some speakers are more efficient than others and require less power to drive them. The manufacturers of most speaker systems make specific recommendations for power amplifiers.

4. For the electronic section of your sound system, you can depend heavily upon specifications. (It is still useful, however, to compare various models using the A-B test.) Some specifications may be more important for your considerations. For example, if a tuner is to be used in an urban area with many strong FM signals to choose from, the selectivity is more important than the sensitivity.

5. When all your components have been chosen, ask the salesperson to connect your selected components together for a comprehensive audition. If there are any incompatibilities, they will show up in this audition. Listen for hum levels, background noise, and any other disturbances. Tune in several FM stations and assess reception and ease of tuning.

The purchase procedure above is a skeletal approach. You should fill in details appropriate to your specific needs and desires. May your new system bring you full enjoyment.

For Further Thought

1. Design a sound system that you would classify as an acceptable "budget" system.

2. Design a sound system that you would like to adorn your living room. By as specific as you are able and justify your choice.

3. Identify factors that would influence your choice of an AM/FM tuner.

4. Would you prefer a receiver or individual components? Why?

Electronic Music

Igor Stravinsky's *Sacre du Printemps* premiered in Paris in 1913. When Conductor Pierre Monteux turned to face the auditorium at the conclusion of the concert, he encountered the ranting howls of a furious audience. Police had to be summoned to quell the tumult. Now we accept the hammering rhythms and dischordant sounds characteristic of *Sacre du Printemps* with alacrity.

In 1900, a new science and a new art arose in Western Europe. Albert Einstein and Paul Klee, both born in 1879, were exact contemporaries. Schönberg was five years their senior. As scientists smashed the atom after upsetting the foundations of classical physics, painters smashed traditional forms and composers repudiated established musical frameworks.

The artist is the mirror of his time. In the twentieth century machines and electronics have had an impact on nearly every aspect of our lives. It is natural to be uneasy with the new. It is to be expected that electronics will be viewed as a dehumanization of music. Yet, has electronics made a fundamental difference? Actually, the fundamental change occurred when *homo sapiens* ceased using the voice alone as the source of musical sound. When Orpheus picked up his lyre, he was using a machine. Contemporary symphony orchestras are really sound factories. Today's electronic devices are merely a new means for producing sounds and stand squarely in the tradition rooted in the distant past.

So new is electronics that the word electronic does not appear in any dictionary prior to 1940. The musical electronic age in America began on October 28, 1952, at New York's Museum of Modern Art. On that day Leopold Stokowski gave a concert. The scene was unusual, for in the middle of a naked stage sat a lone loudspeaker mounted in a simple cabinet.

265

Robert Moog, a pioneer in the development of electronic sound sources, has given us a definition of electronic music:

> Electronic music consists of electronically generated sounds and natural sounds that are modified electronically, assembled into music by magnetic tape manipulation, or performed alive. Electronic music is a medium of expression, not a specific type of music. In fact, electronic music includes music as diverse as Carlos and Folkman's *Switched-On Bach*, the newest Coca-Cola commercial, some of the currently popular rock'n'roll, and representative material from all camps of the musical avant garde.

In their time Mozart and Beethoven were fascinated by the possibilities inherent in the "new" pianoforte that displaced the older harpsichord. They exploited these possibilities. Without doubt, there are composers today who consider the synthesizer as a fascinating new instrument. Whatever they accomplish with this new means, we shall not forget Chopin. Richard Strauss did not make us forget Wagner. Rather, the contemporary artist will exploit the synthesizer, and its sounds will become a part of our musical tradition. It is quite likely that future generations will feel as comfortable with the synthesizer as the current one feels with the conventional instruments of musical sound.

17 Music With the Synthesizer

We started our subject with the audacious claim: all musical sounds can be represented by a curve. With tongue-in-cheek, I suggested that when Mahler composed his *Third Symphony* he was actually designing a curve. By adjustments to the curve, a composer can achieve changes in pitch, loudness, and timbre. How did Mahler realize the particular effects he sought? By bringing together many instruments and assigning to each one a specific role.

With a symphony orchestra a wide range of timbres, pitches, and sounds are possible. For any individual instrument, however, the accessible range is quite limited. The timbre of a trumpet cannot be changed from note to note and the range of frequencies it can cover is a mere fraction of the audio frequency range. (**Note.** A performer can change the timbre of his trumpet with a mute.) Furthermore, once a note has been sounded, its sound can be only slightly influenced.

An electronic music synthesizer is extremely versatile. In principle, any known musical sound can be reproduced by electronic means; in addition, entirely new musical sounds can be created. Its range of frequency transcends the scope of human audibility. It is probably fair to say that the synthesizer is the most expressive musical instrument ever created; however, it is also clear that the Artur-Rubinstein-of-the-synthesizer has not yet appeared. Furthermore, it is still very uncertain what role electronically produced sounds will play in the evolution of our cultural tradition in music.

BASIC COMPONENTS OF AN ELECTRONIC SYNTHESIZER

In the production of sound by a synthesizer any acoustical effect can be reproduced and controlled by electronic means. Among the principal electronic devices employed to accomplish these ends are oscillators, filters, amplifiers, and envelope generators. We shall now consider each of these components individually.

The Oscillators

An oscillator is an electronic device that generates a periodic signal. On a synthesizer, the frequency can be varied from the subsonic range (0.03 Hz) to frequencies of the order of 10,000 Hz. Thus, the frequencies of greatest musical significance are embraced.

Several waveforms can be generated. These include:

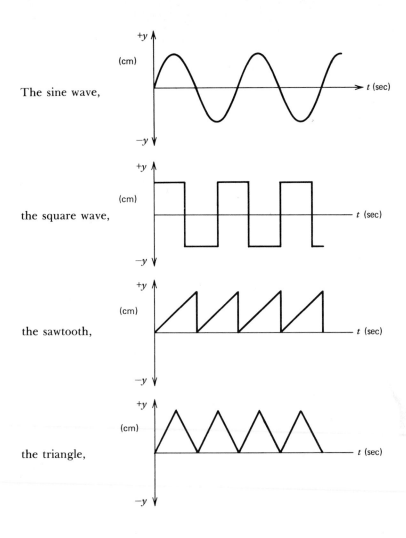

The sine wave,

the square wave,

the sawtooth,

the triangle,

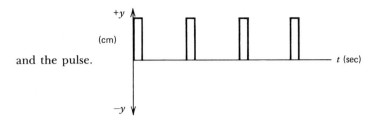

and the pulse.

The width of the pulse is typically variable.

The Filters There are several types of filters used in a synthesizer. These include the low-pass filter, the high-pass filter, the band-pass filter, and the band-reject filter.

A low-pass filter is designed to pass all frequencies up to the cut-off frequency of the filter. Frequencies higher than the cut-off frequency are rejected. The diagram below represents the ideal low-pass filter. Note that in this ideal case, the filter passes 100% of all those frequencies below cut-off and 0% of those frequencies above cut-off. In practice, low-pass filters have a roll-off; that is, they pass all frequencies up to the cut-off frequency and then roll off more gradually. An actual low-pass filter (with roll-off) is shown in the next illustration.

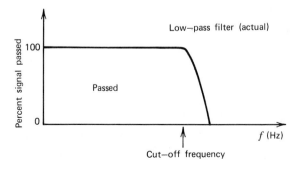

As you might have guessed, a high-pass filter passes all frequencies *above* the cut-off frequency and rejects all lower frequencies. A graph of a high-pass filter is shown below. Finally, a band-pass filter passes a small band of frequencies and rejects all others; in contrast, a band-reject filter rejects a small band of frequencies and passes all other frequencies. Both filters are illustrated below.

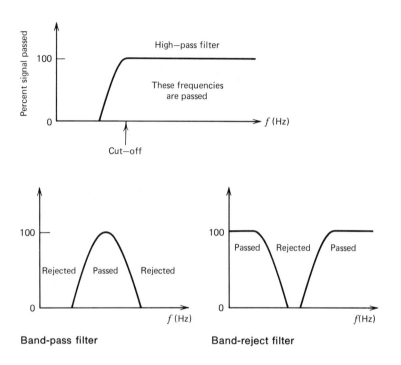

Band-pass filter Band-reject filter

The Amplifier

As we have discussed already, amplifiers enable control to be exercised over the output amplitude.

The Envelope Generators

With envelope generators the attack and release characteristics of a tone can be controlled. There are two types of envelope generators worth considering. The first is called an AR envelope generator and has two adjustable parameters. See the illustrated station. It is apparent that an AR envelope generator provides control over *Attack* and *Release* of the tone.

The second envelope generator is the ADSR type and is more versatile because four parameters, *Attack*, *Decay*, *Sustain*, and *Release*, can be independently varied and thereby give a desired shape to the synthesizer sound. The diagram below shows an envelope with the variable parameters indicated.

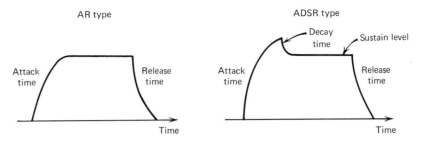

Envelope generators

Envelope generators take advantage of the obvious differences in the way different instruments achieve their characteristic sounds. Obviously, we must start with a waveform characteristic of a particular instrument. But this alone is not sufficient. A flute, for example, is characterized by an attack during which the tone gradually builds up—hence, the "breathy" quality of the flute. The guitar, on the other hand, has a very sudden percussive attack. Thus, the flutelike and the guitarlike waveforms must be shaped if the synthesizer is to faithfully reproduce the sounds we identify with specific instruments. The shaping is achieved by controlling the electronic signal with an envelope generator.

ADDITIVE AND SUBTRACTIVE SYNTHESIS

If we had a large number of oscillators, each capable of producing a pure tone whose frequency could be varied, we could synthesize *any* waveform merely by adjusting the relative amplitudes of the appropriate frequencies. With this method, called additive synthesis, one can be guaranteed success; however, a large array of oscillators is required.

Subtractive synthesis begins with various waveforms that already consist of a large number of pure tones. Let us consider the waveforms discussed earlier and take note of their harmonic contents. In the figure on page 272, the harmonic content of each waveform is shown. A sine wave is harmonically pure and consists only of a fundamental. A square wave contains only odd harmonics of the fundamental. The intensities of these harmonics decrease rapidly with frequency.

The sawtooth waveform contains all harmonics with intensities decreasing quite rapidly with frequency. The pulse, like the sawtooth, contains all frequencies; however, the relative intensities of the harmonics decrease more slowly. The narrower the pulse, the more slowly the relative intensities of the harmonics decrease.

Now let's make use of a low-pass filter for the purpose of subtractive synthesis. We shall begin with a 250 Hz square wave and pass it through

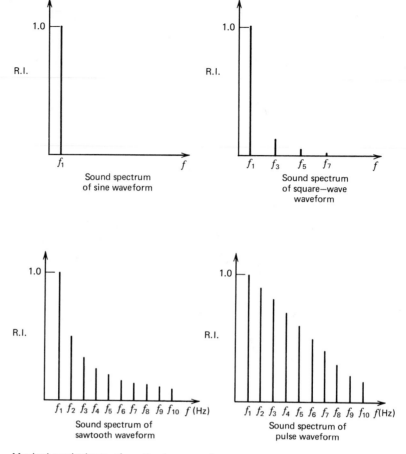

Sound spectrum
of sine waveform

Sound spectrum
of square—wave
waveform

Sound spectrum of
sawtooth waveform

Sound spectrum of
pulse waveform

Musical equivalents of synthesizer waveforms

a low-pass filter whose cut-off frequency is 10,000 Hz. What gets through the filter? In this case, the fundamental and all harmonics up to the 39th ($f_{39} = 9750$ Hz) get through the filter [see part (a) in the following illustration]. If, now, the cut-off frequency is slowly decreased, fewer harmonics get through the filter; that is, higher harmonics are "subtracted out." [See parts (b) to (d).] Finally, when the cut-off frequency of the filter reaches 600 Hz, all harmonics except the fundamental are filtered out and only a 250 Hz sine wave remains. Hence, a harmonic-rich waveform can be modified by filtering. This filtering process is called subtractive synthesis.

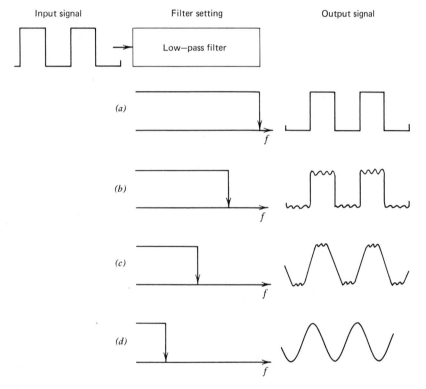

Subtractive synthesis by means of a low-pass filter

PUTTING THE COMPONENTS TOGETHER: SAMPLE PATCHES

What are the basic components of a musical instrument? First, an excitation mechanism of some kind—for example, edge tone, bowing, and lip vibration. Second, a resonator that efficiently transforms the vibrational energy of the oscillating device into acoustical energy. Third, a mechanism for varying the pitch. These components are shown in the diagram on page 274 for a clarinet.

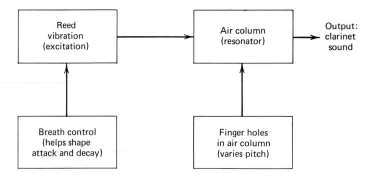

The clarinet symbolically

Each of the components above have their analogues among the components of a synthesizer (see below) and with these components, a clarinetlike sound can be generated.

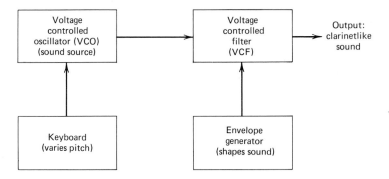

The above diagram represents an oscillator connected to a filter; that is, the signal from an oscillator is fed into a filter. The output of the filter is clarinetlike sound. Such an interconnection of components is called a *patch*. "Setting up a patch" means connecting components together. A patch thereby provides conducting paths for the electrical signals that eventually are transformed into an audio output signal.

A distinction is commonly made between signal paths and control paths. The horizontal lines above represent signal paths and the vertical lines are control paths. Sound can be generated by a synthesizer if only signal paths are employed. For example, sound would result from the patch shown on opposite page.

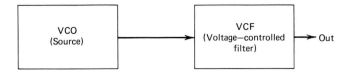

Or, if we want to control the amplitude of the output signal, the patch would have the following form:

Additional patches involving only signal paths are illustrated below.

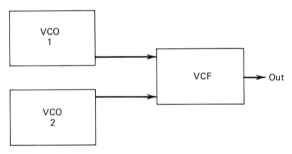

Here two oscillators are simultaneously fed into the filter. The two oscillators can be tuned to the same pitch or to some desired musical interval.

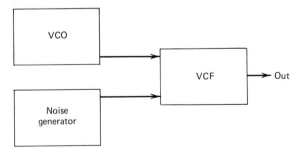

A noise generator has been added to accompany an oscillator.

Obviously, it is important to do more than just generate sound. The objective is to be able to control all factors leading to a desired resultant sound. To do this, control voltages are used. Consider the patch next

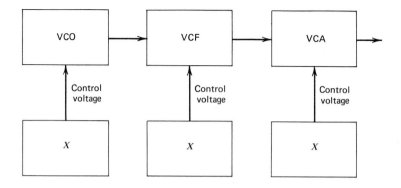

where the voltage controls are symbolized by an *X*. One of the most popular sources of voltage control is a keyboard. If the keyboard controls an oscillator, a chromatic scale can be played. As successive keys are depressed, an incremental voltage is applied to the oscillator that raises the pitch by an incremental amount. As shown below, with 1 volt we can play 12 tones; that is, there is 1 volt per octave. The keyboard has the major advantage of being able to create an instantaneous change in the output of the controlled device as soon as an instantaneous change in the control voltage is made.

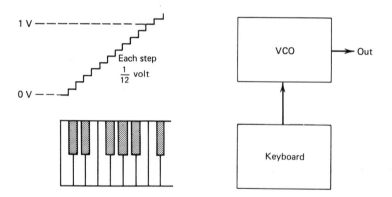

We shall illustrate only a few of the many patches that use control voltages. You can imagine many others. Examine the patch on opposite page. Assume the cut-off of the filter is set at maximum so that *all* frequencies of the noise generator get through. How do you imagine the output sounds? The resultant sound can simulate a "choo-choo" train. How is this done? If the control oscillator is set at a frequency of 2 Hz, the amplifier is instructed to amplify the signal from the noise generator twice each second with the resulting "choo...choo." Since the sound output will follow the sine curve waveform, the "choo" will

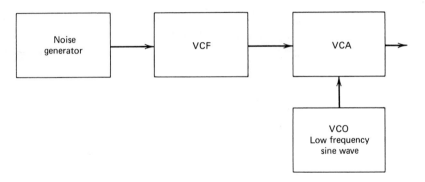

build up and fall away in intensity. A sine wave control voltage can also be used to achieve nice vibrato effects, as shown in the patch below.

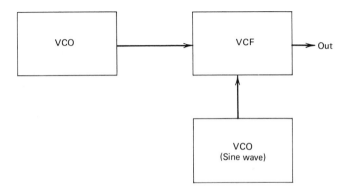

The attack and decay of a tone is controlled by the ADSR control. The patch below has vibrato as well as the facility to shape the resulting sound.

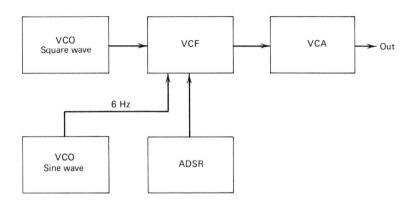

These are just a few examples of how the various components work together to create sound. When you get your hands on a synthesizer, you will become very accustomed to creating the patches necessary to achieve the desired result.

For Further Thought

1. A broad band of frequencies form the input to a low-pass filter. The output of the low-pass filter than leads a high-pass filter. Let us assume that each filter is ideal. If the low-pass filter has a cut-off frequency of 2000 Hz and the high-pass filter a cut-off frequency of 1800 Hz, what is the output of the combination?

2. The combination of filters above are equivalent to what type of filter?

3. Design a band-pass filter to give frequencies in the range of 1200 to 1500 Hz. To do this design work you have a low-pass and a high-pass filter whose cut-off frequencies are variable.

4. A 320 Hz pulse passes through a low-pass filter with a cut-off frequency of 2500 Hz. Describe the output signal of the filter.

Answers to Selected Problems

Chapter 2

3. 440 Hz
5. 5.33 newtons

Chapter 3

3. 2.64 m
4. 11,000 m/sec
5. (a) 2 watts/m^2 (b) 0.222 watt/m^2
6. 70 db
7. 89.77 db
8. By factor 1.73

Chapter 4

1. 77.8 db
2. 60 phons
3. Approximately 48 phons
4. 100 phons
5. (a) 30 phons (b) Approximately 40 phons
6. 4 sones
7. 74 phons

Chapter 6

1. (a) 0.00446 sec (b) 0.125 sec
2. (a) 0.00463 sec (b) 0.125 sec
3. (a) 0.00228 sec (b) 0.00114 sec, 0.00076 sec, 0.00057 sec, etc.
5. 720 m/sec
6. (a) nodes (b) node

Chapter 7

1. (a) 2 Hz (b) 4 Hz (c) yes
2. (a) 446 Hz or 434 Hz (b) Can't tell
3. Approximately 1075 Hz
7. (a) Approximately 40 Hz (b) Approximately 140 Hz
8. $f_{C1} = 147$ Hz, $f_{C2} = 293$ Hz, $f_{C3} = 146$ Hz
10. 0.00909 sec
11. 12.25 Hz

Chapter 8

3. 49.9 newtons
4. 138 m/sec
5. 276 Hz
6. (a) 20.8 cm (b) 10.45 cm
7. 10,144 Hz
9. (a) Semiclosed (b) 0.43 meter
10. 160.8 Hz

Chapter 9

1. (a) 1st, 2nd, 4th, 5th, 7th, etc.
2. (a) 0.00101 kg/m (b) 166.7 Hz, 333.4 Hz, 500 Hz, 833.3 Hz
3. Multiples of the fourth and fifth harmonics
4. $\frac{4}{5}$ of each cycle
6. $0.8T$

Chapter 11

2. (a) Approximately 285 Hz (b) 0.0193 m^3
5. Flute: 6-holes, clarinet: 11-holes
7. (a) 129 Hz (b) 59 cm
9. 35.4 cm
13. 0.011 sec

Chapter 12

3. 1.02 sabines
4. 14.99 sabines (at 1000 Hz)

5. 7.8 db
6. 57.5 Hz, 86.3 Hz, 115 Hz
7. 1.54 sec (at 1000 Hz)
8. 0.26 sec (at 1000 Hz)
9. (a) 8.83 sec

Chapter 13

2. Interval 2
3. Most consonant: fourth 3 Least consonant: fourth 1

Chapter 14

4. 20 db
7. 10^7
8. (a) AM (b) 268 m
15. 138 Hz and higher

Index